NATAL
The Garden Colony

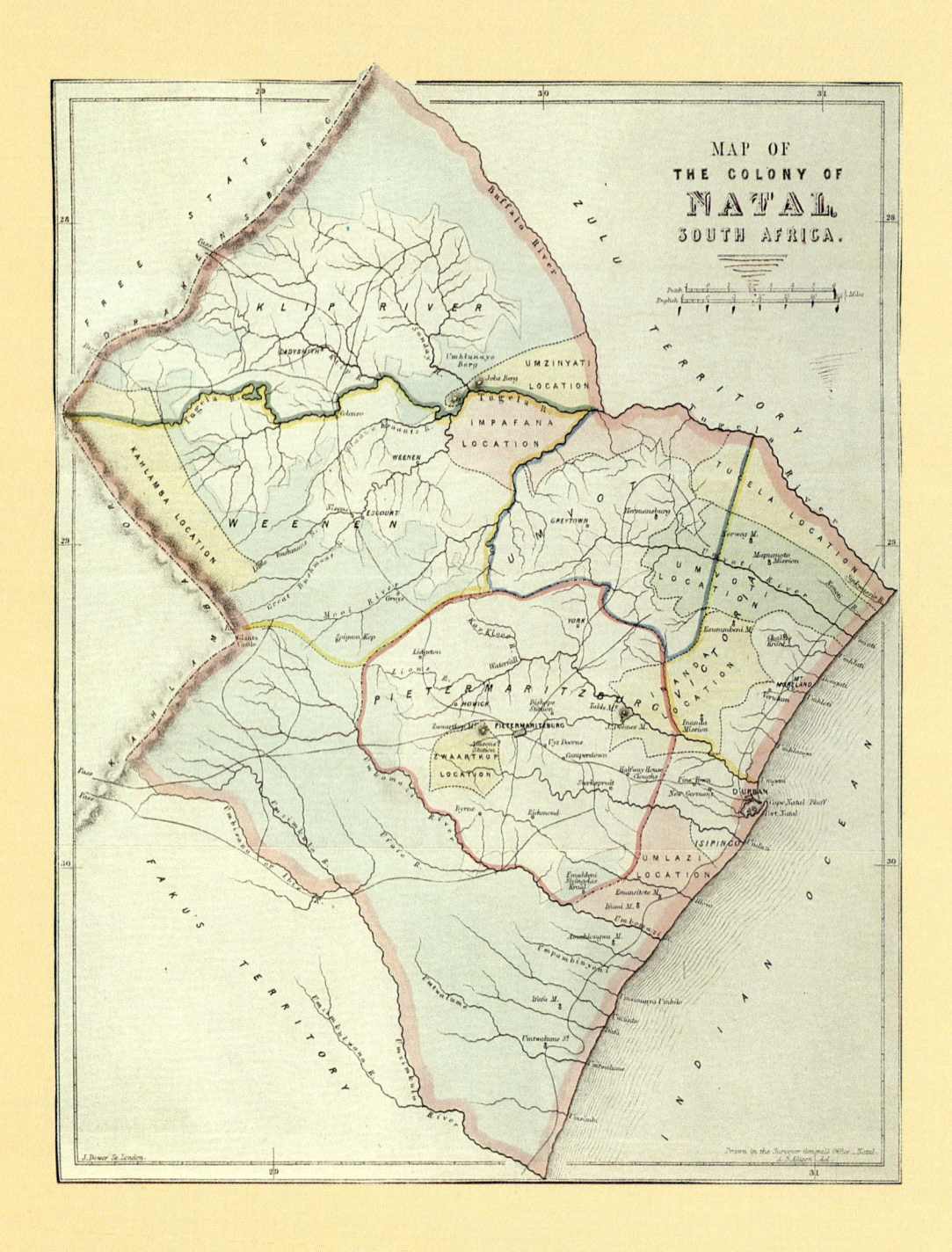

Map of the Colony of Natal, 1859. This interesting map of Natal is contained in Dr Robert Mann's book on the colony which appeared in 1859 and which was aimed at attracting new settlers. Dr Mann was Natal's first superintendent of education.

NATAL
The Garden Colony

Victorian Natal and the Royal Botanical Gardens, Kew

Donal P. McCracken and Patricia A. McCracken

FRANDSEN PUBLISHERS
Sandton
1990

Published by
Frandsen Publishers (Pty) Ltd
P O Box 122 · FOURWAYS · 2055
South Africa

© **Donal P McCracken** & **Patricia A McCracken**

Designed by **Robin Frandsen**
Reproduction by **Pointset**, Randburg
Printed and bound by
National Book Printers, Cape Town

This book is copyright under the Berne Convention in terms of the Copyright Act (Act 98 of 1978). No part of this book may be reproduced or transmitted in any form or by any means electronic or mechanical including photocopying, recording or by any information storage and retrieval system, without the permission in writing from the publisher

ISBN 0-9583124-1-9

This volume is dedicated to
Gerard and Evelyn McGuire

Contents

Foreword ix

List of illustrations x

Abbreviations and notes on terminology xi

Preface xii

Acknowledgements xiv

Introduction xv

Chapters :

1. Sir William Hooker and the colony of Natal 2
2. Early plant exchange 20
3. Imperial Kew and the heyday of Natal botany 27

The paintings of Marianne North 34

4. Botanical institutions develop 44
5. The Indian summer 60
6. Later plant exchange 78

Select bibliography 88

Index 94

Foreword

SIR GEORGE TAYLOR

D.Sc., F.R.S., Ll.D., F.R.S.E., F.L.S., V.M.H.

Director of the Royal Botanic Gardens, Kew, 1956-1971

The influence of the Royal Botanic Gardens, Kew, during the expansive Victorian Empire which coincided with the auspicious development of Kew under the régime of the Hookers, father and son, pervades this work. Exhaustive study of the Kew archives has produced a critical and most informative volume giving a powerful insight into the activities which established the fame of Kew throughout the world as a centre of botanical endeavour in the dissemination of economic and ornamental plants, the accumulation of material as the basis for colonial floras, the publication of numerous fundamental works, and the training of men who went overseas to develop new botanical gardens throughout the British Empire. These men received solid help from Kew but there were bountiful reciprocal benefits for the parent establishment in the enrichment of the home collections both of living plants and of herbarium specimens.

This book is a case study of one such relationship between Kew and the settlers of one of the colonies. Natal was remote, subtropical, impoverished and dangerous for settlers because of malaria and the occasionally hostile Africans. Despite its tiny settler population, colonial Natal remained staunchly pro-British. Furthermore, it was soon realised that it was a botanical treasure house which had been largely unexplored before the 1850s.

The volume succinctly chronicles the achievements of a succession of enthusiastic amateur botanists and others who introduced economic and ornamental plants into the colony and in return sent to Kew a stream of Natal species of all groups. According to the records about half of the wardian cases - at the time the accepted means of transporting plants by sea through the tropics - received by Kew from Africa came from Natal.

The husband and wife team - she as a professional journalist, he as historian - have produced a unique presentation of a sphere of activities which cannot fail to interest the historian, the botanist, the gardener and the conservationist and all who enjoy accounts of human enterprise. It reveals a host of interesting episodes which would otherwise remain obscure in the Kew archives and I value an opportunity to commend this work.

List of illustrations

Durban from Mr Currie's residence (*front cover*)
Haemanthes insignis (*back cover*)
MN 354, White convolvulus & Erythrinas (*title page*)
Map of the Natal colony, 1859 iv
The Great Palm House, Kew xv
The herbarium and library, Kew xvii
Sir William Jackson Hooker xix
The Orangery, Kew xx
The interior of the Great Palm House, Kew xx
The pond and Museum 1, Kew xxi
Mark McKen 3
Encephalartos villosus 4
A page from Robert Plant's diary 5
Stangeria paradoxa 6
Streptocarpus saundersii 7
Stapelia gigantea 8
William Tyrer Gerrard 9
Streptocarpus polyanthus 10
Plant hunters encampment, northern Natal 12
Sandersonia aurantiaca 13
Dr Peter Cormac Sutherland 15
Hunting for animals and plants 17
Bishop John William Colenso 18
Haemanthus insignis 22
A wardian case 22
Thunbergia natalensis 24
Hypoxis latifolia 25
Interior of the Orangery 25
Opening pages of the Kew plant exchange book 26
Sir Joseph Dalton Hooker 29
River scene, Victoria lilies 30
Frontispiece of *Curtis's Botanical Magazine* 31
Albuca nelsoni 32
Katharine Saunders 33
Marianne North's paintings 34-39
Haemanthus katherinae 40
Letter from Thomas Baines to Sir Joseph Hooker 42
Thomas Baines' sketch of *Aloe bainesii* 43
Anemone fanninii 47
Disa cooperi 48
Julius Wilhelm (William) Keit 49
Haemanthus katherinae 50
Littonia modesta 51
Note from Robert Topham to Sir Joseph Hooker 55
Durban from Henry William Curries residence 57
John Medley Wood 58
Plan of the Durban botanic gardens 59
Carriages journeying to the gardens 59
Sir William Thiselton-Dyer 61
Heliophila scandens 62
Eucomis bicolor 63
Richardia (Zantedeschia) rehmannii 65
Pietermaritzburg botanic gardens 66
Sir Theophilus Shepstone 67
Interior of the Jubilee conservatory, Durban 68
The colonial herbarium, Durban 68
A note from John Medley Wood to Sir William. Thistelton-Dyer 69
Kniphofia modesta 70
Maurice S. Evans 71
Nymphaea coerulea (stellata) 72
The Jubilee conservatory, Durban 75
Alberta magna 81
Aloe bainesii 81

Abbreviations and notes on terminology

The following abbreviations have been used:

Allan - Mea Allan, *The Hookers of Kew, 1785-1911,* (London, 1911)

Bean - W. J. Bean, *The Royal Botanic Gardens, Kew: Historical and Descriptive,* (London, 1908)

Britten and Boulger - James Britten and G.S. Boulger, 'Biographical index of British and Irish botanists', *Journal of Botany,* vols 26-9, (1888-91)

D.L.H.M. - Durban Local History Museum

Gunn and Codd - Mary Gunn and L.E. Codd, *Botanical Exploration of Southern Africa,* (Cape Town, 1981)

Huxley - L. Huxley, *Life and letters of Sir J.D. Hooker, based on materials collected and arranged by Lady Hooker,* (London, 1918), 2 vols

K.A. - Kew Archive (see note below)

Kew Bulletin - *Royal Gardens, Kew: Bulletin of Miscellaneous Information,* (1887 onwards)

Kew Report - *Report on the progress and condition of the Royal Gardens at Kew,* (to 1865 by Sir W.J. Hooker; from 1866 - 82 by Sir J.D. Hooker; no reports appeared for 1883 - 86 after which the *Kew Bulletin* was published)

K.C.A.L. - Killie Campbell Africana Library University of Natal, Durban

UNLESS OTHERWISE STATED, letters quoted from the Kew Archives are referred to in footnotes merely by the writer's name and the date of the communication. If the date falls between 1844 and 1858 the letter may be found in the African letters volume LIX, and if between 1859 and 1864 the letter comes from the African letters volume LX. For the years 1865 to 1900 Natal letters can be found in the three South African letter volumes depending on the surname of the writer: A-G, volume 189; H-N, volume 190; and P-W, volume 191.

Throughout the text we have used the term 'botanic gardens' rather than 'botanic garden'. There are two reasons for this: first that is what these institutions in Durban and Pietermaritzburg called themselves in the Victorian era, and second because Victorian curators recognised the desirability of the Linnean idea of the institutions they controlled being made up of a collection of several specialist gardens within the whole. We have not, however, succumbed to the Victorian propensity for the overuse of capital letters and on this thorny subject we have been guided by the eminent historians, A.F. Pollard and T.W. Moody.

Finally, some purists may query our not consistently referring to John Medley Wood with the full surname 'Medley Wood'. The family name was in fact Wood, his letters at Kew are filed under Wood and his early publication (1879), *Classification of Ferns,* is imprinted under the name John M. Wood. He came to use the name Medley Wood and where appropriate we have used this.

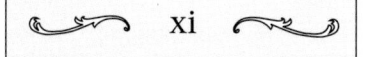

Preface

NO CASE STUDY HAS EVER been attempted of the relationship and inter-dependence of the world-famous Royal Botanic Gardens at Kew outside London with a colony in the Victorian empire. This volume aims to rectify the omission. The colony of Natal is an excellent example to survey. Its existence as a colony falls roughly within the Victorian era, its settlers were often people of some education, who were interested in the flora around them and took a pride in working with Kew. Finally Natal, with its three distinct major ecological regions, proved a paradise for the Victorian plant hunter.

A semi-tropical coastal belt contained a plethora of luxuriant species, including ferns, orchids, euphorbias, and thick impenetrable bush. Trees were on the whole of small to medium height, with a few exceptions such as the Natal mahogany *(Trichilia dregeana)*, which could grow to over 90 feet (30 metres) in height. Numerous bays and lagoons provided the swampy conditions where mangroves proliferated.

Inland lay a meandering escarpment with a more temperate climate and subject to perennial mists. This area was in parts dense with yellowwood forest *(Podocarpus spp.)*, intermingled with ironwood, stinkwood and sneezewood, and marked the beginning of the undulating midland plateau with an elevation of between about 2 000 and 5 000 feet (600 and 1 500 metres). A major feature of this plateau was thornveld acacia in an open grassland region with an abundance of aloes in parts. The midlands also contained such collectable plant species as cycads, ferns, helichrysums and orchids.

Finally the highlands rose from about 5 000 feet (1 500 metres) to the great peaks of the Drakensberg range with Basutoland (now Lesotho) beyond. Here, as in the midland escarpment, forest was to be found. While most of Natal's ericas grew only in this region, many of the midland flower and woody species were also to be seen here.

So different was the Natal coastal flora from that of the two higher regions and so abundant was the flora in all three areas that Kew was not slow to take an interest in Natal and in acquiring botanical specimens from the region. Kew, in turn, assisted the struggling colony. Robert Topham's expression of appreciation, sent in August 1872 to Kew's director, was well-merited praise:

How many will have to thank your thoughtfulness in Natal in years to come. My experience is many promise and do nothing. You promise and do promptly.

Whether it was seeds and plants of possible agricultural worth to the colony, horticultural specimens, timber trees, or just ornamental garden, street and park species, Kew supplied the wants of the colonists free of charge.

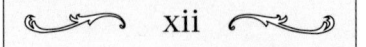

If any criticism is to be levelled at the many Natal plant collectors, it is that they were too enthusiastic. In January 1873 William Keit, the curator of the Durban botanic gardens, complained that the area around the town had 'been well nigh stripped of what was worth taking'. Ross's *Flora of Natal* lists some 45 species, most of them named in the colonial period, whose localities are now unknown as they are now rare or possibly extinct in Natal. However, it can be forcefully argued that population growth with all its ensuing consequences - the expansion of Natal due to the demand for wattles for building and for plank and post timber - was far more significant in its impact than a collector wandering over the veld with a bag in his hand. Moreover, these early botanists created an awareness of the flora of the region which well compensated for any disruption they might possibly have caused to the environment.

In the writing of this book we have faced the problem of identifying both scientific and popular plant names which are now regarded as archaic. Having worked through the Plant list and Exchange volumes in the Kew archive we are fully aware of the truth of Sir William Hooker's observation:

The same name is applied to several trees in one colony, and to others in other colonies; and these names are often purely arbitrary (applied by memory, or originated in a whim, or in an erroneous idea of the tree to which they are given) are soon lost sight of and often wholly forgotten.

We have tried to make use of the modern accepted nomenclature and have mentioned in the text when we have given the contemporary Victorian name. Despite the fact that handwriting and spelling vary greatly in the sources, and given that taxonomy is constantly developing, we trust that we have been as accurate as possible.

Acknowledgements

WE SHOULD LIKE TO TAKE THIS opportunity to thank the various institutions and individuals who have helped us in the preparation of this book. The Killie Campbell Africana Library of the University of Natal has been most encouraging about this project. In particular Mrs J. Simpson was invaluable for her knowledge of relevant manuscript material and her helpful disposition. Equally encouraging was the archive of the Royal Botanic Gardens, Kew. Miss Sylvia FitzGerald, the chief librarian and archivist, gave us access to whatever material we requested from the archive and Kewensia rooms as well as allowing us the freedom of Kew's superb library. The photocopying of well over a thousand documents was done by her staff with remarkable efficiency. Miss Leonora Thompson reacted with patience and good humour to the many demands we made of her. We thank her for this. Thanks also to Mr B. Spencer and the staff of the Don Africana Library, Durban.

In addition we should like to thank the late Mr R.G. Strey, whose work on botanical history will assist scholars for many generations to come; Ms Susyn Andrews; Mr Don Bain; Dr Peter Brain; Mr John Chartres; Mr W.J. Forgrave; Mrs D. Fourie and Mrs Marie Jordaan of the Botanical Research Institute, Pretoria; Mr Gren Lucas, Deputy Director of Kew; Keeper of the Kew Herbarium; Emeritus Professor J.L. McCracken; Mr Brian Schrire, former South African liaison officer at Kew; Mrs S.O'B. Spencer and Sir George Taylor.

We are grateful to the following who have generously sponsored the publication of this volume: the Stanley Smith Horticultural Trust of Britain and the United States of America, and the South African Sugar Association.

Our thanks go to our publishers, Frandsen Publishers of Sandton, who have spared little in an effort to produce a fine publication.

Of the 848 paintings which were produced by Marianne North during her extensive travels, only 16 were executed during her stay in Natal. It was felt that this work would be incomplete should we not reproduce all of these Natal paintings. Consequently the rights for reproduction have been obtained from Kew Gardens and these works are reproduced within the pages of this book.

Only one of these 16 paintings have ever been reproduced before and this reproduction is to be found in *A Vision of Eden: The Life and Work of Marianne North* (London 1980). We acknowledge the publishers, Messrs Webb & Bower of Exeter. This painting is entitled 'A Remnant of the Past near Verulam' (Catalogue 383).

Introduction

THE ROYAL BOTANIC GARDENS, KEW, was a product of the eighteenth-century movement known as the enlightenment. The quest for rational knowledge and natural beauty found one of its most enduring expressions in the development of scientific botany. In such an age of patronage it was fortunate that this development had the active support of many of the ruling families in Europe, who were prepared to encourage the establishment of botanical institutions.

The present gardens at Kew are in essence the amalgamation of part of the Richmond (or Ormonde) Lodge estate with the smaller Kew (or White) House estate. Richmond Lodge was bought in about 1721 by George II when he was Prince of Wales. It was to remain a royal residence for the next half-century. The grounds became notorious for the extraordinarily shaped follies created by George II's wife, Queen Caroline, the most grotesque of which was called Merlin's Cave.[1] When George II died in 1760, the estate passed to his grandson and heir, George III, who maintained an establishment there until 1772.

The Kew estate adjoined the Richmond estate and ran from the present southern boundary of the gardens, roughly between what is now the Holly Walk and Kew Road up to Kew Green. Thus on its southern side and most of its western side it was bordered by the Richmond estate. In 1730 Kew House was leased to George II's reckless son, Frederick, Prince of Wales, and his wife, Princess Augusta. It is said that the writer Alexander Pope gave Frederick a dog which he kept at Kew House; on its collar was inscribed the couplet:
I am His Highness's dog at Kew
Pray tell me, sir, whose dog are you?[2]

Why Frederick chose to live so close to the father he detested is not known but for the future development of Kew it proved a most fortuitous decision.

In the seventeenth century the grounds of Kew had already been noted for their beauty. While Frederick was not entirely lacking in interest in such matters it was Princess Augusta who aimed to develop part of the estate into a scientific establishment. When Frederick died in 1751, before his father, the Dowager Princess

The Great Palm House, Kew
Erected between 1844 and 1848, the Great Palm House was the symbol of Kew's greatness and of Victorian 'botanical imperialism' Its erection was made possible by the intervention of the process of making curvilinear glass.
(From, *The Gardeners' Chronicle,* 5 August 1876).

1 An illustration of this strange folly can be seen in Bean, p.5.
2 T.M. Martin and A.R. Hope Moncrieff, *Kew Gardens,* (London, 1908), p.15.

The scientific heart of Kew - the herbarium (background) **and library** Once the house of the King of Hanover, in the early 1850s it was placed in the hands of Sir William Hooker, the director of Kew. Wings were added in 1877 and 1902. It was here that a small staff received numerous botanical specimens, mainly from the empire. They also carried on a voluminous correspondence with colonial and foreign curators. The correspondence from Natal was received here, where it is still housed.
(From, Bean's *The Royal Botanic Gardens, Kew*).

Augusta, debarred from ever becoming queen, devoted much of her energies to the estate. She was greatly helped in this by the much maligned earl of Bute and by her head gardener, William Aiton. Three major developments occurred under Princess Augusta's patronage: the grounds were improved and ornamented with a number of follies designed by Sir William Chambers; a hothouse known as the great stove was constructed; and about 1759 a small area of some nine acres (3,5ha) was set aside for a botanic garden. This was to remain a private royal institution for the next 80 years.

In 1772 the dowager princess died and her eldest son, George III, took possession of the Kew estate which he subsequently purchased. As George III already owned Richmond Lodge, he was able to move his establishment from there to Kew House and have the former demolished. About that time he had a small picturesque summerhouse retreat, the Queen's Cottage,[3] built near the site of the old Richmond Lodge. The Richmond estate had already been landscaped by Capability Brown under George III's instructions. Queen Caroline's follies had been removed and what is now the Hollow or Rhododendron Walk excavated. The two very attractive estates were not formally joined until a public right of way, called Love Lane, about where Holly Walk lies today, was closed in 1765.

In 1802 Kew House was demolished and the adjacent and smaller Kew Palace or Dutch House was used as the royal residence until 1818. It then lay empty till 1899. A scheme to erect a new royal palace in medieval style near the site of Kew House reached the stage of an outer shell being constructed, but the king's mental illness meant that further building ceased. The new Kew Palace or 'Bastille' remained a curious and enormous folly, a short distance from the Dutch House and the orangery, for a quarter of a century until it was demolished in the late 1820s, by which time the heyday of

[3] The Queen's Cottage is named after George III's wife, Queen Charlotte.

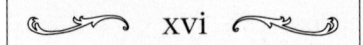

the Georgian garden had long passed.

When George III first moved to Kew House he had been on the throne for 12 years. His mental health had not yet seriously deteriorated nor had the disaster of the American revolution yet occurred. The luckless Bute had passed from royal favour and the king strove with great zeal to rid politics of ruling oligarchic cliques and reassert the position of the crown. His championing of Sir Joseph Banks and support for Banks in asserting the importance of Princess Augusta's old botanic garden may be seen as part of this process. A scientific royal botanic garden under an eminent botanist, importing plants from remote and exotic regions and rivalling the Chelsea Physic Garden and the Parisian Jardin des Plantes, could only enhance the prestige of the crown.

From 1772 to 1820 Banks, acting as the unofficial director, first with the support of William Aiton and then from 1793 with that of his son William Townsend Aiton, established Kew's international reputation. When Francis Masson was despatched to collect at the Cape of Good Hope in 1772 he was the first of many plant hunters to risk their lives in the service of Kew.

The catastrophic decline in Kew's fortunes over the 30 years between 1810 and 1840 may simply be attributed to lack of royal support. The prince regent and future George IV and his brother, the future William IV, looked elsewhere for their aesthetic pleasures. As the Chelsea garden fell on bad times after Phillip Miller's death in 1771, so the passing of George III and Joseph Banks in 1820 accelerated the decline of Kew.

By the time George III's granddaughter, the Princess Victoria, ascended the throne in 1837 the very existence of the botanic garden at Kew lay in the balance. A royal commission consisting of Joseph Paxton and John Wilson under the chairmanship of the botanist Dr John Lindley was appointed to investigate Kew. The Lindley report, presented in March 1838, recommended that the botanic garden should be reconstituted as a national scientific institution. Should this not be done, Lindley proposed that it should then be abandoned. The commission was actively supported by the duke of Bedford who made it his life's endeavour to save Kew.

The lord steward, however, would have none of this. Not only did he propose turning the hothouses over to practical horticultural purposes but he also went so far as to offer Kew plants to two other botanic bodies. The intervention of the earl of Aberdeen on the side of the scientists decided the issue and the small walled botanic garden was transferred to the government department of the commissioner of woods and forests on 1 April 1840.[4]

In 1841 Sir William Jackson Hooker (1785-1865), the eminent professor of botany at Glasgow university, was appointed director of Kew. The decision to appoint Hooker was as enlightened as the one to save Kew as a public scientific institution. In his character, his scientific achievement and his drive to establish Kew as Europe's leading botanic gardens, Hooker proved the ideal director. Though he was not of the patrician mould of Banks, Hooker with his personal charm was able to achieve far more for Kew than had his illustrious and overbearing predecessor. Within five years all but the royal residence and several segments of the old Kew estate had been transferred to Hooker's control. The rather cramped and run-down old botanic garden had the advantage of possessing the pleasure grounds, an attractive stretch of land which had already been laid out in parts and on which several of Princess Augusta's more pleasing follies had

4 Hadfield, pp.315-6.

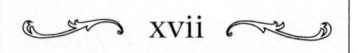

Sir William Jackson Hooker (1785-1865) Sir William Hooker was director of Kew from 1841 until his death in 1865. More than any other man he built up Kew to become the world's foremost botanical institution. It was Hooker's encouragement that prompted some early Natal colonists to take up plant hunting. A polished gentleman, Sir William was an influential figure in mid-Victorian London.
(Courtesy of the Royal Botanic Gardens, Kew).

---

5 Ray Desmond, 'The historical setting of Kew', in F. Nigel Hepper (ed), *Royal Botanic Gardens, Kew: Gardens for science and pleasure,* (H.M.S.O.,1982), p.13.

6 Hadfield, pp.317-28

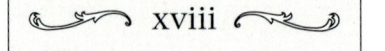

---

survived. These included the pagoda, the orangery and the ruined arch. It was an excellent foundation on which W.A. Nesfield could landscape the gardens with spectacular vistas and walks.[5]

The grandeur of Burton and Turner's great palm house, completed in 1848, symbolises the importance of the new Kew. Though the palm house might contain some plants dating from Banks' time, such as the *Encephalartos longifolius* specimen brought back from South Africa by Masson in 1775, most of the plants it was to contain were newly collected in the tropics and subtropics. From as early as 1843, Hooker recommended the practice of sending Kew collectors to remote parts of the globe.[6] He also en-

The Orangery, Kew Relic of the pre-Victorian age. The Orangery, built in 1761 by Sir William Chambers for Augusta, princess dowager of Wales, was archaic by the time Kew was revived in the 1940s. It was difficult to grow plants successfully because the design limited the amount of light entering the building. It was therefore turned into a wood museum. It now houses a book and gift shop.
(From, Bean's *Royal Botanic Gardens, Kew*).

The pond and Museum 1, Kew This museum of economic botany was erected in 1857. It was enlarged in 1881 to facilitate the housing of a large collection of Indian botanical specimens. Many objects from Natal were also exhibited in this building.
(From, Bean's *Royal Botanic Gardens, Kew*).

couraged freelance collectors and settlers in British colonies to send him specimens. The latter were often fired by Hooker's enthusiasm into establishing colonial botanic gardens or stations in their respective colonies. In Australia in the 1840s and 1850s, for example, moves were taken by colonists to found such institutions in Hobart, Melbourne and Brisbane. Closely connected with this imperial venture was the establishment in 1847 of the world's first museum of economic botany at Kew.

In a remarkably short time Sir William Hooker had not only created a gracious and botanically interesting garden but he had also forged a scien-

The interior of the Great Palm House, Kew Tens of thousands of visitors flocked here annually to enter a magical world of semi jungle. For the Victorian Natalians there was the satisfaction of seeing old friends from the colony's flora.
(From, *The Gardeners' Chronicle*, 5 August 1876)

7 W.F.Monypenny and G.E. Buckle, *Benjamin Disraeli*, (London, 1914), III.385.

8 E.H.Brookes and C. de B. Webb, *A history of Natal*, (University of Natal Press, 1965), pp.17-20.

9 A survey of Natal newspapers in the early 1850s will show how concerned the colonists were that Natal might be vacated by the British authorities.

10 K.A.: Bowker, 21 April 1885.

tific link between Kew and a network of imperial botanical institutions. The resulting benefit for the advancement of botanical knowledge as well as for the economic development of the empire through the exchange of plants must be cited as one of the most positive aspects of nineteenth century imperialism.

The colony of Natal

WHEN DISRAELI, WRITING TO Lord Malmesbury in August 1853, described the British north American colonies as 'a millstone around our necks', he would have been more accurate so to single out Natal.[7] Lying between the Umzimkulu and Tugela rivers, Natal was annexed by the British in 1843 for strategic reasons. Since the 1820s a small band of hardy British traders visited by the occasional missionary had lived in primitive conditions in a swampy bay area called Port Natal, on the site of the present city of Durban. They hunted the elephant, hippopotamus, lion and antelope which abounded in the area for ivory, skins and hides. They also traded with the Nguni who had lived in the region for some considerable time and with whom they had a relationship which fluctuated between peaceful coexistence and open hostility.[8]

The arrival in 1837 of the Boer voortrekkers, self-imposed exiles from the Cape colony, introduced an effective colonising element to Natal. The prospect of the Boers defeating the Zulus in 1839 emphasised this and heralded the consolidation of what was for a brief time called the Republic of Natalia, some 250 miles (400 kilometres) north of the Cape colony's eastern frontier. A Boer republic with access to the sea and located close to a powerful African community was unacceptable to the British authorities in London. After a brief military encounter with the Boers at Port Natal, Britain annexed the republic.

Though many Boers trekked back over the Drakensberg mountains into the interior, the British felt it expedient for the moment to remain in possession of the region, attaching it until 1856 as a district of the Cape Colony, with its own lieutenant governor. That Natal was not abandoned, as many believed it would be and as the Orange River Sovereignty was later to be, was a momentous and costly decision.[9] Since it lacked extensive exploitable resources, Natal was bypassed by the great waves of emigration from the British Isles from the 1840s onwards. A Natal botanist rightly commented to Kew in 1885, 'Natal like Ireland is not sought after as the abode of the Anglo-Saxon'.[10] Despite the efforts of one major immigration scheme (the Byrne settlement of 1849 to 1852), the settler population grew at a painfully slow rate and was overwhelmingly outnumbered by the Nguni.

The Economic development was

xx

equally slow in its growth and there was a series of depressions compounded by a number of serious droughts. A labour problem, only partly solved by the import of Indian workers, and a troublesome sand bar across the entrance of Durban harbour, further exacerbated Natal's economic woes.[11] In addition throughout its history the very existence of the colony was threatened from without and within. Early settlers faced periodic raiding from the surviving earlier inhabitants of the region, the Khoisan. More serious was the threat from the Nguni, many of whom had been confined to rural locations. There were African revolts in 1857, 1870 and 1906 and a major war (the Anglo-Zulu war) in 1879. Furthermore the colony was briefly invaded by the Boers in 1881 (the first Anglo-Boer war) and suffered partial occupation during the much more serious second Anglo-Boer war (1899 to 1902).

That the colonists in this most patriotically British colony should devote so much energy to botany is an interesting phenomenon. One might have supposed that the only tangible link in Natal with the world of natural science in Britain would be the segment of Joseph Paxton's London 1851 great exhibition building which served as a sugar warehouse in Durban.[12] But the settlers' combination of pride in their new settlement and love for their old homeland was a powerful force in many colonies. There was a desire to recreate many of the institutions they had left behind. Proper botanic gardens were a luxury most colonies could not afford but they were a feature of the age in Britain and Kew was also actively encouraging their establishment. Then colonial rivalry came into play: if Queensland had a botanic gardens then New South Wales must have one too.

In the South African context the old Dutch East India Company garden at the Cape, which had become a mere pleasure park by the early nineteenth century, was designated a botanic gardens in 1849. Grahamstown founded its botanic gardens a year later. In 1865 a botanic gardens was laid out at King William's Town. In the 1870s others followed suit. Fort Beaufort, Port Elizabeth, Kimberley, Graaff-Reinet and Queenstown, among others, established what they termed botanic gardens. Pretoria in the Transvaal Republic had a botanic gardens in the 1870s and, as will be seen, Natal possessed two such institutions.[13]

But the South African gardens, unlike their counterparts in Australia or India, were starved of money and many soon deteriorated into commercial nurseries or public parks. It is an indication of the success of Kew in promoting botanical awareness in the young colony of Natal that the Durban botanic gardens was the only such establishment before the foundation of the National Botanic Gardens of South Africa at Kirstenbosch in 1913 to maintain its integrity as a scientific institution.

11 For the economic history of colonial Natal, see Bill Guest and John M. Sellers, *Enterprise and exploitation in a Victorian colony: Aspects of the economic and social history of colonial Natal*, (University of Natal Press, 1985)

12 George Russell, *The history of old Durban*, (Durban 1899), p.93.

13 See Donal P. McCracken and Eileen M. McCracken, *The way to Kirstenbosch*, (N.B.G., 1988)

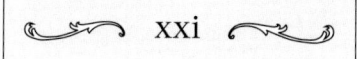

Chapter 1

Sir William Hooker and the colony of Natal

WHEN NATAL WAS ANNEXED to the British empire in 1843 little was known of the colony's flora. From the 1820s to the early 1840s a few plant collectors had visited the region. Among them were James Bowie, W.F. Drège, Miss M.C. Owen, Ferdinand Krauss and possibly C.H. Whendemann. Natal did not gain its first resident botanist until 1841 when the eccentric Wilhelm Gueinzius settled in Durban, as Port Natal had been renamed. Gueinzius' links were, however, with his patron, Professor E.F. Poeppig of Leipzig, and not with Kew. While Sir William Hooker was aware of Gueinzius' activities, he did not receive any of the specimens which Gueinzius sent to Europe. Only occasionally did Kew receive material from Natal; nearly 500 of Krauss' specimens are housed in the Kew herbarium, as is an *Argyrolobium ascendens* from Drège that Kew described as a 'miserable scrap'.[1]

Despite the efforts of these early collectors, their achievements were somewhat limited and well into the 1840s Kew regarded Natal as being remote and mysterious. This is graphically illustrated, when in 1844, whilst discussing the possible origins of *Limonium peregrinum (Statice rytidophylla)*, both Sir William Hooker and the celebrated Irish botanist William Harvey agreed that seeds of the plant in their possession were more likely to have originated from the Cape than from 'so distant a region as Port Natal'.[2]

The forging of a link between Kew and Natal came about because of two new developments. The first was the establishment by a number of enthusiastic and prominent colonists of a Natal Agricultural and Horticultural Society in Durban in April 1848. After encountering considerable difficulties, in June 1851 the society managed to get possession of 25 acres (10 hectares) of crown land for the purpose of establishing a garden to experiment with exotic and indigenous plants which might be of economic and commercial value to the colony.[3] The site for the garden was attractive but inaccessible, being two miles (3,2 kilometres) to the east of the village of Durban, on the lower slopes of an aeolian ridge known as the Berea and across a swampy and mosquito-infested vlei. In 1854 the area of the garden was doubled, thus giving the society a plot of 50 acres (20 hectares), more than it could manage.[4]

The second development was an official emigration scheme from Britain to Natal, organised by Joseph Byrne between 1849 and 1852. In March 1850 the *Illustrated London News* noted:

Great exertion is making at this moment to direct the great stream of emigration towards

---

1 For accounts of these collectors, see their respective entries in Gunn and Codd; A.W. Bayer, 'Aspects of Natal's botanical history', *S.A.J.S.*, 67, 8 (August 1971), pp.401-11; and E. Percy Phillips, 'A brief historical sketch of the development of botanical science in South Africa and the contribution of South Africa to botany', *S.A.J.S.*, 27, (November 1930), pp.39-80. See also B.J.T. Leverton, *Records of Natal (1823-1828)*, (Pretoria, 1984), pp.42-5.

2 *Curtis's Botanical Magazine*, (1844), t4055. See also, *Hooker's Journal of Botany*, 1, (1849), p.32.

3 D.L.H.M., Minute book of the Natal Agricultural and Horticultural Society, 1848-1854, H1130A.

4 The 1854 title deeds are housed in the Durban City Estates Department, file TC15/5J636C.

the new colony of Natal, which may be described as a long strip of fertile country in South Africa, along the coast and inland as far as the Drakensberg Mountains, its port lying 1 000 miles to the north-east of the Cape of Good Hope...Iron, coal, copper and plumbago [graphite] are found in the colony; and there is abundance of wood of every description. The Dutch colonists are hospitable, religious and moral; and the Zoolus, a native tribe living in Natal, are intelligent, docile, and shrewd.[5]

In all about 5 000 Byrne settlers arrived in the underdeveloped colony, among them a number of plant collectors who had been encouraged to emigrate to Natal by Kew.

The professional collectors

THE MOST PROMINENT OF the professional plant collectors at this time were McKen, Plant and Gerrard. Unlike the epic venturing collectors such as Purdie in tropical America, Burke in north-western America or Joseph Hooker in northern India, these Natal collectors were essentially freelance: though given the official blessing and encouragement of Kew, they were paid only by results. While an interesting collection of plant specimens might be expected to elicit between £30 and £50 from Kew, many collectors found it necessary to diversify into more lucrative fields of collecting. Though the large mammal capture market was dominated by others, the Natal plant collector could find ready overseas sale of insects, birdskins, small reptiles, live birds (especially parrots) and even shells. Robert Plant openly admitted to William Hooker that despite his devotion to botany he found that 'insects are more portable and permit me to add more profitable'.[6] In several instances collectors were forced to take up other fulltime employment. Both McKen and Plant at various times were curators of the horticultural society's garden, a circumstance which gave rise to much ill-feeling as they were often accused of neglecting the garden in favour of their private collecting activities.

MARK JOHNSTON McKEN arrived in the colony on the *Emily* in October 1850.[7] He had managed the Golden Grove sugar estate at Port Morant in Jamaica from 1840 until 1849. In August 1848 he wrote to Kew, sending botanical specimens for the Kew museum and asking Sir William Hooker to keep him in mind for botanical work which might be available. Sixteen months later he was still importuning Hooker for a preferment. Failure to be selected as a collector on a Scottish botanical venture to Oregon finally decided McKen that he must sail for Natal and find employment growing sugar or cotton.[8] Hooker was encouraging and provided McKen with a large wardian case containing 29 species of possible economic viability for the new colony,[9] which McKen handed over to

Mark McKen (1823-72) McKen was the greatest of the early colonial plant hunters in Natal. He was curator of the Durban botanic gardens from 1851-53 and again from 1860-72. This fiery Scot alienated some settlers with his eccentric ways but he was always on excellent terms with Sir William Hooker at Kew, whom he supplied with many plants from Natal and Zululand.
(Courtesy: National Botanical Institute, Pretoria)

---

5 *Illustrated London News*, 16 March 1850.

6 K.A.: Plant, 15 April 1857.

7 For details on McKen, see Britten and Boulger, 27, (1889), p.340; Gunn and Codd, p.238; and B.D. Jackson, *Guide to the literature of Botany*, (London, 1881), pp.350-1.

8 For McKen's early correspondence with Kew, see K.A.: South American Letters, 1841-1851, vol LXX, 4 August 1848, 19 December 1849, 18 March 1850 and 24 June 1850.

9 K.A.: Exchange book, outward, (1848-1859), 6 July 1850.

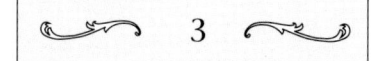

ENCEPHALARTOS VILLOSUS This fairly well-known Natal cycad appears to have been introduced into Europe in the 1860s. It was described by Professor Charles A. Lemaire of Paris in 1867. For a number of years it was known as *E. mackenii*, after Mark McKen. *(Curtis's Botanical Magazine*, T6654, [1882])

the horticultural society on his arrival in Durban. In June 1851 McKen was appointed curator of the society's garden on the meagre salary of £50 a year plus a hut. He was, however, allowed for his own benefit any excess produce from the gardens which remained after subscribers had been given plants. Despite hostility between the society's committee and McKen, he settled well into his new position and was soon collecting and sending plants and dried specimens to Kew. In November 1851 he wrote to Hooker:

I occupy the whole of my leisure time drying plants and I have now the pleasure in informing [you] that my herbarium is rapidly increasing in size, specimens of the whole are prepared for you and shall go with a collection of live plants and seeds by a vessel to sail duly from this Port to London in February.[10]

McKen remained at the gardens until August 1853 when he left to manage what was to become the Saunders' sugar estate at nearby Tongaat. While there he often ventured into Zululand to trade goods and collect plants. He found his employer's family interested in botany and in particular he did much to inspire Katharine Saunders' interest in the subject.[11] In December 1860 he returned to his former position as curator of the Durban botanic gardens and remained at this post until his death in 1872.

McKen was not always in tune with the more refined denizens of this small and poor settlement, though many referred to him as "the professor". He was a heavy drinker and he neglected to lay out or indeed clear most of the garden properly. Nor did he label the plants. He also favoured with plants only those subscribers of whom he approved. McKen did, however, do much for the agricultural development of the colony with his experiments in growing such crops as arrowroot and sugar cane, the latter eventually becoming the backbone of the region's economy. As a plant collector he was surpassed only by Gerrard. McKen took the trouble to learn and write down the Zulu names for various plant species and sent specimens to Kew for 20 years. In return for such contributions the Hookers provided the colony with important garden and crop species. Kew also often named some of the Natal novelties after their discoverers. By the 1890s McKen's name was commemorated in the genus *Mackenia* and in a number of species. These include the orchid *Eulophidium mackenii*, the lily *Eriospermum mackenii* and the gourd *Peponia mackenii*.[12] Most of the species bearing McKen's name were described by John Baker of the Kew herbarium. As, however, with many other species named in the Victorian period, developments in taxonomy have led to the reclassification of some of McKen's plants, including the cycad *Encephalartos mackenii* - this older name is now regarded as a synonym for *E. villosus*.

McKen collected mainly within a radius of some 20 miles (30 kilometres) of Durban. Only occasionally did he venture further afield to the moun-

---

10 K.A.: McKen, 28 November 1851.

11 R.G.T. Watson, *Tongaati: An African experience*, (London, 1960), chapter 3.

12 See J. Medley Wood, *Preliminary catalogue of indigenous Natal plants*, (Durban, 1894).

John Sanderson (1820 - 81) this newspaper editor was for over a decade president of the Natal Agricultural and Horticultural Society, in which position he was often in friction with his fellow Scot, Mark McKen, the curator of the Durban botanic gardens. Lieutenant Govenor Keate said that Sanderson had the character of seldom agreeing with anybody about anything. Sanderson was a keen amateur botanist who in later life specialised in studying orchids. He was a frequent correspondent with Kew. (Courtesy: Killie Campbell Africana Library)

tainous region of the Noodsberg north-east of Pietermaritzburg. But Robert Plant, who arrived in the colony after McKen, was more adventurous in this respect and Sir William Hooker described him as 'a zealous naturalist and able collector'. Plant had first made a name for himself as an experimental nurseryman at Cheadle in England and as compiler of a *New Gardeners' Dictionary*.[13] These enterprises brought him into contact with the English naturalist William Wilson Saunders, who specialised in collecting African flora. Why Plant decided to emigrate to Natal is not known; about the same time his brother, N. Plant, became a plant collector in South America.[14] Saunders arranged with Hooker that the great systematist George Bentham would receive, name and distribute specimens sent by Plant from Natal. Bentham lived and worked at Kew and although not on the institution's payroll he was very much part of its scientific side. He worked on several colonial floras as well as collaborating with Joseph Hooker on the great systematic work on flowering plants, *Genera Plantarum*, commenced in 1857. From 1861 to 1874 Bentham was president of the Linnean Society. He eventually donated his library to Kew. That someone so eminent should undertake to act for Plant is a reflection of the light in which Kew regarded Natal in the 1850s. Plant also maintained a link with the famous Chelsea Physic Garden in the city of London.[15]

Plant arrived in Natal with high hopes and was soon setting off north to Zululand. He despatched an ox wagon ahead of him and with a large number of oxen and African helpers he crossed the Tugela and began collecting plants, insects and shells. In an account of the expedition Plant mentions the magnificent mpande palms and the numerous and beautiful terrestrial orchids of Zululand. The expedition, however, was not a success. Plant failed to find his wagon at a pre-arranged meeting place

13 *Paxton's Botanical Dictionary*, (London, 1868 edition), p.446.

14 *Hooker's Journal of Botany*, III, (1853), p.125.

15 Britten and Boulger, 28, (1890), p.153; and Gunn and Codd, p.282.

STANGERIA PARADOXA This interesting Natal plant is now known as *S. eriopus*. For a number of years there was some doubt about whether it was a fern or a cycad. It was discovered by the eccentric German naturalist William Gueinzius, who lived near Pinetown in a house also occupied by three African rock pythons. He gave Natal's surveyor general, Dr Stanger, a specimen of this plant which he took to London in 1851. In turn Stanger gave the specimen to the celebrated Nathaniel Ward who had it planted in the Chelsea Physic Garden. In 1854 Stanger sent another specimen to Kew. By then the plant had already been described in Germany as a *Lomaria*. It was now reclassified by Dr David Moore of Glasnevin botanic gardens as a new genus which he named in honour of Dr Stanger. In November 1855 Kew was greatly pleased when Robert Plant sent them a box containing eight 'fine stems' of the *Stangeria*. (*Curtis's Botanical Magazine*, T5121, [1859])

---

16 R.W. Plant, 'Notice of an excursion in the Zulu country', *Hooker's Journal of Botany*, IV, (1852), pp.257-65.

17 K.C.A.L.: Notebook of R.W. Plant's visit to Madagascar; and K.A.: Sanderson, 8 April 1854.

18 K.A.: Plant, 26 April 1856.

19 *Natal Mercury*, 26 April 1856

20 K.A.: Exchange book, inward, (1848-1859), 19 May 1855.

21 *ibid.*, 23 November 1855.

22 K.A.: Plant, 18 April 1856.

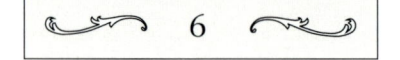

6

and he lost many of his oxen through sickness and because he had to kill some for food. Though he had collected deep into central Zululand, he was forced to jettison most of the specimens he had found because of lack of package space and because he inadvertently stumbled into a war waged by the Zulu king Mpande. Though Plant sent a consignment of bulbs to London on his return to Durban in February 1852, the eight-month expedition produced little more than an interesting and colourful account of the trip which was published in *Hooker's Journal of Botany* in 1833.[16] It is claimed that this was the first botanical paper by a resident of Natal to appear in a scientific journal.

Though disheartened by the trip, Plant was determined to continue collecting even though this meant his venturing far up the coast to east Africa and to some of the islands in the western Indian Ocean.[17] In 1854 he returned to Zululand to collect but by August of that year circumstances forced him to follow McKen's example and become curator of the horticultural society's garden at an annual salary of £60. Like McKen, he found that his employers regarded his private collecting unsympathetically and the society's committee restricted him to only one week's collecting a year.[18] None the less Plant tried to put some order into the hotchpotch of a garden left by McKen and in April 1856 was rewarded when the *Natal Mercury* described the garden (which lay to the east of Durban) as that 'Eastern paradise'.[19]

In May 1855 Plant returned a wardian case to Kew which they had originally sent to the garden in August 1854. Unfortunately many of the species which he packed into the case died en route to Britain and those which survived (four species of fig, two of jasmine, three laurel-like shrubs and one specimen each of white pear, gardenia and yew) were described by Kew as 'none of much interest'.[20] But a box that Kew received from Plant in November 1855 proved much more exciting. It contained eight 'fine stems' of the cycad-like *Stangeria eriopus* (*S. paradoxa*).[21] In a letter to Sir William Hooker in April 1856 Plant noted how dissimilar a plant looked in its 'native wilderness' from how it appeared in 'a flowerpot in England'. He also discussed the problems of identifying plants in Natal, where many names might exist for a single species. He concluded by expressing his delight at hearing that Kew was so far in advance of all other botanic gardens.[22]

By this time Plant was heartily fed up with the horticultural society. As curator he was expected to organise exhibitions and annual shows as well as put the garden in order and supply subscribers with plants. He was advised by Wilson Saunders to give up his post and employ his time more profitably. Consequently Plant resigned and established himself as a farmer at Vaal Hoek near Tongaat. Here he became the pioneer of tea growing in the colony. Sanderson described this farm as:

a delightful place, a little tongue of land between two streams where he was naturalizing ferns from all quarters, besides planting coffee, tea, arrowroot, etc.[23]

However, Plant could not resist collecting whenever he was free and even helped the horticultural society in this field. Then, to Plant's great joy, Sir William Hooker wrote to him in January 1857 suggesting the possibility of Plant working on a *Flora Natalensis*, to which he replied:

I am so wedded to this beautiful colony that nothing would give me greater pleasure than to assist in making its richness known and if your leisure will admit to a work of such magnitude I would beg to be reckoned as an aid so far as my humble abilities may be useful.[24]

He then explained the necessity of going over the same area at different seasons to be sure of finding as many different plants as possible. So unassuming had Plant been in agreeing to undertake the large-scale collection, preparation and despatch of dried specimens to Kew that before he received a reply from Sir William, in a letter to him when sending some ferns and some *Streptocarpus* plants whose species he did not know, Plant reaffirmed his desire to work on the flora, 'I beg to place myself unconditionally at your disposal for the purpose named'.[25]

With renewed enthusiasm Plant set about collecting as many different plants as he could find, which he dried and stored at his farm. Then in late 1857, leaving his wife and five children at home, he set off on an epic

STREPTOCARPUS SAUNDERSII This member of the Gesneriaceae family was sent to the noted English plantsman W. Wilson Saunders by Robert Plant in the 1850s. Saunders sent it on to Kew, where it was named in honour of him, Saunders, who did much to propagate and sell South African plants. (*Curtis's Botanical Magazine,* T5251, [1861])

collecting trip north through Zululand and into Portuguese East Africa, taking a wagon and a number of African helpers. In 1852 Plant had written of the St Lucia area:

Elephants seem in great plenty all over this district, as we frequently saw herds of them. There are but few inhabitants of this part, which argues but little for its healthiness.[26]

His words proved to be prophetic for in mid-March 1858, whilst he was

23 K.A.: Sanderson, 31 October 1858.

24 K.A.: Plant, 15 April 1857.

25 K.A.: Plant, 26 May 1857.

26 Plant, 'Excursion in the Zulu country', p.264.

STAPELIA GIGANTEA Though the largest flowering and best-known plant of the family Stapelieae, this plant, which was discovered in Zululand in the early 1850s by Robert Plant, was slow to gain recognition. When Plant died in 1858 his widow sent it to the Durban botanic gardens. After two years there it finally flowered and two years later, in 1862, the curator Mark McKen sent a specimen to Kew. Curiously, it was to be another 15 years before it was described by N.E.Brown. The plant was also collected in Natal by William Gerrard. (*Curtis's Botanical Magazine*, T7068, [1889])

returning from his second Zululand expedition, Plant fell ill and died of fever at St Lucia. The servants buried him there and returned to the farm with their master's belongings and the plants he had collected.

Though Mrs Plant was left unprovided for, she succeeded in considerably developing the farm despite the colonial government's refusing to grant her a gratuity. In 1862 she had the satisfaction of having the tea produced on her estate shown at the London colonial exhibition. She also supplied James Brickhill, one of the future tea-growers, with his first 10 plants.[27]

In June 1858 Mrs Plant wrote a rather pathetic letter to Sir William Hooker thanking him for the kindness he had shown her husband and offering to send Kew what dried plants she could find about her home. What live plants there were she had sent to the horticultural society's garden in Durban.[28] One of these proved to be the

'greatest' of the *Stapelia*. In December 1853 Rev. Edward Armitage, who was briefly visiting the colony, mentioned in the lecture on Natal botany which he delivered in Pietermaritzburg:

Mr Plant found a flower of this kind [Stapelia] in the Zulu country as large as the top of a hat, and calculated to inspire as much horror by its appearance as the sight of a large serpent might be supposed to occasion.[29]

This plant flowered in the horticultural society's garden in August 1860 but it was only in 1862 that McKen sent a specimen of the giant plant to Kew, stating that it came from the 'Umvelos river in Zulu country'. Surprisingly it was another 15 years before N.E. Brown of Kew named the plant *Stapelia gigantea*.[30] Plant's name was commemorated in another *Stapelia*, *S. plantii*, in the orange-flowered *Gloriosa plantii* and in the fern *Lastrea plantii*. That so few plants bear his

27 K.A.: James Brickhill, October 1886.

28 K.A.: Mrs Plant, 29 June 1858; and Sanderson, 31 October 1858.

29 Rev. E. Armitage, 'Some observations on the botany of Natal', in J. Chapman, *Travels in the interior of South Africa*, (London, 1868), II.463.

30 K.A.: McKen, 30 August 1862.

name is a sad reflection on the ill fortune which beset his endeavours.

BY THE EARLY 1860s Sir William Hooker was energetically fulfilling his position as director of Kew, now assisted by a larger staff. On the scientific side they included his son, Joseph Hooker, the assistant director, and Professor Daniel Oliver who was soon to be appointed keeper of the Kew herbarium and library, a position he was to hold for over a quarter of a century. Sir William Hooker also retained the services of the celebrated botanical artist Walter Fitch, who drew and coloured the magnificent illustrations in *Curtis's Botanical Magazine*, which Sir William edited from the 1820s. In addition Sir George Bentham was always ready to give his expert advice and help on botanical matters.

This Kew establishment was well aware of the lack of a Natal flora. In the 1868 edition of William Harvey's *Genera of South African Plants*, which was first published in 1838, Joseph Hooker pointed out that the work had originally been written before the botanical exploration of Natal.[31] Harvey's *Thesaurus capensis*, published in 1859, lists few Natal plants and volume one of his and Sonder's celebrated *Flora Capensis*, which came out a year later, contains the admission, 'Our flora...presents little more than an outline sketch of the Northern and North Eastern Regions and of the Natal Colony'.[32]

With the death of Plant in 1858, Kew's hope of a *Flora Natalensis* faded. It now became clear that Natal would have to be properly incorporated into Harvey and Sonders' *Flora Capensis*. Though neither author was on the Kew staff, their flora was part of a Kew scheme, devised in 1863, to produce 12 colonial floras which were to be 'scientific, yet intelligible to people of ordinary education'. They were to cover Australia, southern Africa, British north America, the British West Indies, New Zealand, Ceylon, Hong Kong, Mauritius and the Seychelles, British Guiana, Honduras, British west Africa and British India. Later Professor Oliver undertook a flora of tropical Africa. Of the 12 regions designated for study, it was estimated that only three had more than 3 000 plant species: India had 12 000, southern Africa 10 000 and the Australian colonies, including Tasmania, 8 000 species.[33] By 1860 Sir George Bentham had completed the Hong Kong flora and Dr Joseph Hooker and Dr Thomson had published, at their own expense, volume one of *Flora Indica*. Volume one of Harvey's work was seen as the first of ten volumes. Harvey was professor of botany at Trinity College, Dublin, and had spent some time at the Cape as colonial treasurer before returning home to Ireland in 1842 because of ill health. He had collected for a flora while at the Cape so had a knowledge of the terrain, though he was largely

William Tyrer Gerrard (d. c.1866)
Gerrard was a collector in the style of the classical plant hunters such as Bowie and Cunningham. He collected in south-eastern Africa as well as on the islands off east Africa. He discovered over 150 new species and is especially remembered for the new genera, *Gerrardanthus* and *Gerrardina*. Gerrard is one of two martyrs of botany associated with Natal. He died on Madagascar. (Courtesy: National Botanical Institute, Pretoria)

31 W.H. Harvey, *The Genera of South African Plants*, (London, 1868 edition), p.6.

32 W.H. Harvey and O.W. Sonder, *Flora Capensis*, (Dublin, 1859-60), I.7.

33 K.A.: Kewensia room, Colonial flora volume.

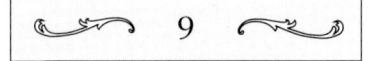

STREPTOCARPUS POLYANTHUS
This attractive little perennial is to be found in damp sheltered spots. It is a member of the Gesneriaceae family. In 1853 Captain Garden brought Kew some ferns. From among the roots of these ferns there appeared seedlings of this plant. Though John Sanderson had sent the Kew herbarium a dried specimen of this species, these were the first live plants Kew was to receive. (*Curtis's Botanical Magazine*, T4850, [1855])

ignorant of Natal. After Plant's death Harvey had a prospectus circulated in Natal to encourage plant collecting.[34] He and the Hookers also decided that for the purpose of the flora they should make use of the services of a professional collector, William Tyrer Gerrard, to collect and dry as many Natal plants as possible. Gerrard was remarkably successful and between 1861 and 1865 managed to cover large tracts of the colony and parts of Zululand.[35] Occasionally he collected with McKen and in 1870, after Gerrard's death, a pamphlet on ferns entitled *Synopsis Filicum Capensium* was published in Pietermaritzburg under the names of these two collectors. The two botanists had got on well together and for a while Gerrard lived in 'the oldest house in Berea', inside the grounds of the botanic gardens.[36]

Where possible Gerrard himself identified the numbered specimens, although on one list that he sent, he admitted, 'I alone am responsible for any blunders in the names from No 1 to 780. Original names and numbers lost.'[37] Gerrard sent most of his specimens to Kew. Those which had not been identified were forwarded to Harvey in Dublin[38] and once Harvey had identified or named them he sent Kew the names. This procedure also applied to specimens sent by some other South African collectors. In March 1872 Kew purchased additional Gerrard specimens, though from whom it is not known. In total Gerrard appears to have sent some 2 000 specimens to Harvey and Kew. It is therefore little wonder that his name is commemorated in two genera, *Gerrardanthus* and *Gerrardina*, and in the names of over 150 species, including many Compositae, Asclepiadaceae, Convolvulaceae, Orchidaceae, Amaryllidaceae, Liliaceae and the krantzberry tree, *Gerrardina foliosa*. Oddly one of the plants he discovered, *Streptocarpus cooksonii*, was not described until 1955.[39]

The early 1860s witnessed one of colonial Natal's frequent economic depressions. The money gained from collecting plants, insects and birds proved inadequate for Gerrard and no alternative employment offered itself: by then McKen was well ensconced, with a salary of £150 a year, in his second term as curator of what was now called the Durban botanic gardens. Thus in March 1865, when Gerrard despatched his latest collection of plants to Kew (for which he received £30), he wrote to Sir William Hooker:

I believe I informed Dr Harvey I had resolved not to collect plants, but it is hardly possible for me to carry out such a resolution - Ferns are too enchanting objects to be left alone.[40]

So Gerrard left Natal, now a sick man, and sailed to Mauritius where he visited the governor, Sir Henry Barkly, an amateur botanist. Gerrard described Barkly as being 'more than a

---

34 K.A.: Sanderson, 31 October 1858.

35 Britten and Boulger, 27, (1889), p.18; *Gardener's Chronicle*, 1866, p.1042; Gunn and Codd, pp.165-6; *Journal of Botany*, 1866, p.367.

36 K.A.: Wood, 24 February 1889.

37 K.A.: Plant list, vol II, f 301.

38 K.A.: Sanderson, 4 July 1864 and 11 October 1864. For an example of this interchange, see *Hooker's Icones Plantarum*, 3rd series, vol II, (London, 1876), p.12.

39 Barbara Everard and Brian D. Morley, *Wild flowers of the world*, (London, 1970), plate 82.

40 K.A.: Gerrard, March 1865.

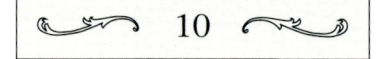

10

father to me'. From Mauritius, Gerrard sailed in May 1865 to the mysterious island of Madagascar to recommence collecting. But he had only a few months to live for he died 'far away from friends and home' at Foul Point, north of Tamatave on the east coast of Madagascar.

His death was lamented in Natal and at Kew. A colonial newspaper deplored the loss of 'so accomplished and indefatigable a naturalist' and Professor Oliver was later to describe Gerrard in the *Icones Plantarum* as 'that excellent collector'. Plant and Gerrard were not the only Kew-backed collectors to become 'martyrs of botany' in this period: in 1860 Charles Barter died on Baikie's Niger expedition and in 1864 Richard Oldham succumbed to dysentery at Amoy in China. Even in 1842, when Bentham had suggested to the young Joseph Hooker that he spend some time collecting in the tropics, Hooker's curt reply had been 'Have not you Botanists killed collectors a-plenty in the Tropics?'[41]

By the mid-1860s the heyday of the professional Kew-backed plant collectors had passed. Indeed it may be claimed that William Tyler Gerrard was the last of the long line of professional collectors sponsored or in some way financially backed by Kew, a line which had begun with Francis Masson, 93 years earlier. No longer was it necessary for Kew to risk the lives of specialist collectors: the proliferation of colonial botanic gardens and botanic stations in correspondence with Kew proved a much more efficient system. Kew also benefited from the activities of collectors working for British nurserymen or for wealthy British private collectors. A good example of how this system operated is the case of Wilson Saunders, who introduced Plant to Kew and who himself supplied Kew with Natal plants from as early as 1856. Though resident in Britain Wilson Saunders employed collectors, the most famous of whom in the Natal context was Thomas Cooper, who married the daughter of the Kew botanist N.E. Brown. Cooper, who had a number of South African plants named in his honour, collected in Natal in 1861 and 1862. In the 1865 *Kew Report* it was noted:

The collection of cacti, aloes, succulents, and bulbs in No 7 [greenhouse] has been for the most part repotted and very greatly improved; and has also been materially increased, chiefly through the liberty of W. Wilson Saunders F.R.S., whose almost annual contributions are both the most valuable and most numerous that this establishment has received since its foundation.[42]

The amateur collectors

THE POPULARITY OF BOTANY in Natal grew rapidly, as is reflected in the substantial number of collectors listed by Harvey in his *Flora Capensis*. Considering the size of the settler population (not much more than 10 000 by then), as well as the fact that the collectors Harvey listed represented only the most ardent botanists, the enthusiasm for the Victorian plant craze in Natal becomes clear.[43] The professional plant collectors were few in number compared to these amateur botanists.

Kew became increasingly dependent on these amateurs but that brought with it major problems. There was a tendency, at times encouraged by Sir William Hooker, for colonists to look only for interesting new species, or novelties as they were termed. Joseph Hooker, however, was worried about novelties and the attention they diverted from botanical research. He even ventured to criticise Harvey: 'You have the old error of preferring novelties to anything else'.[44] Huxley pointed out that 'mere species-mongering' led to terrible confusion, with different names being assigned to the same

41 Huxley, I.167.

42 *Kew Report*, 1865), p.3. See also reports for 1856, 1862, 1871 and 1873.

43 The point is emphasised by Phillips in his 'Historical sketch', p.406.

44 Huxley, I.372, Joseph Hooker to William Harvey, 10 January 1857.

Plant hunters encampment in northern Natal ('Itoli Mountains looking N.N.W. by N.10°, 1852'). Watercolour by Captain Robert Garden. Captain Garden was a professional soldier who spent much of his time when stationed in Natal in the early 1850s plant collecting. Being a man of some standing he was able to hunt animals and collect plants in style as this painting of his wagon train well illustrates. Wagons were regarded as the best vehicles to collect from as they could cope with most terrain; were slow moving so plants could be spotted from them; and were capable of carrying large specimens such as cycads. A wagon, however, could cost £150 to buy or 15 shillings a day to hire. The famous Natal pioneer Henry Francis Fynn, who was not 'fond of bulbs and flowers', angered Garden by his refusal when in the bush to ask Zulus where plants could be found.
(Courtesy: Killie Campbell Africana Library, Campbell Collection of University of Natal, WCP, 930)

plant in different regions. Considerable confusion had arisen, for example, over the correct names of colonial timber on display at the 1851 Crystal Palace exhibition.[45] This problem, which had already been highlighted by Sir William Hooker as has been seen, and which was not unique to Natal, was partially solved by the publication of many of the colonial floras and then, from 1895 onwards, of the famous *Index Kewensis*.

The difficulty went further than the simple duplication of names for a single species: so excited were collectors at finding novelties that often all else was forgotten. Referring to Natal, Professor Oliver noted in *Icones Plantarum* that collectors must give precise information concerning the locality in which a plant was found. The name 'Natal' was not enough. Too often plants were sent to Kew unlabelled.[46]

Although some collectors like Sanderson argued that the discovery of novelties encouraged others to collect,[47] it would be incorrect to condemn all Natal amateur botanists as being only interested in having their names' commemorated by the naming of some colourful lily or orchid. Amateur botanists were very much of two varieties: those who were knowledgeable about botany and those who did not know much but were interested in the subject. As might be expected it was mainly the former who corresponded with and sent specimens to Kew. Most enthusiasts, though, had some links with Kew at one time or another, but these could be indirect as with Dr and Miss Armstrong, for example, who sent specimens to Kew via Dr P.C. Sutherland.

The link between the professional collectors and the amateur enthusiasts was the horticultural society in Durban and, from 1874, the Pietermaritzburg Botanic Society. The first corresponding secretary of the horticultural society, in 1848, was the colonial surveyor-general, Dr William Stanger, who had visited Australia and taken part in the disastrous 1841 Niger expedition before arriving in Natal in February 1845.[48] He became acquainted

45 Allan, p.200.

46 See K.A.: Sanderson, 9 November 1852.

47 K.A.: Sanderson, 3 January 1859.

48 D.L.H.M.: Horticultural society minute book, H1130A, 13 June 1848. See also Britten and Boulger, 28, (1890), p.313; and Gunn and Codd, pp.331-2.

SANDERSONIA AURANTIACA (Sanderson's Christmas bell or the Chinese lantern lily). This attractive lily of the grassveld and mountainous regions of Natal is a member of the Liliaceae family. It was named in honour of John Sanderson, president of the Natal Agricultural and Horticultural Society, who discovered it in November 1851 at Field's Hill near Pinetown and also at Swartkop near Pietermaritzburg when on an expedition into the interior. The plant was described by Sir William Hooker. (*Curtis's Botanical Magazine*, T4716, [1853])

13

with Wilhelm Gueinzius, who gave him a plant which had its leaves pinnately veined and which was considered to be a new genus of the cycad family; in 1851 Stanger took this plant, across to England, to Dr Nathaniel Ward, the inventor of the wardian case. The specimen was planted in the Chelsea physic garden. Three years later Stanger sent Kew a specimen. Though already described in Germany as a *Lomaria*, a fern genus, it was renamed *Stangeria paradoxa* by David Moore of the Glasnevin botanic gardens in Ireland.[49]

Stanger received some exotic plants from Kew and in return sent specimens which included a *Hypoxis*, *Begonia natalensis*, *Streptocarpus grandis* and the now rare *Clivia gardenii*. The last of these was named after Captain Robert, later Major, Garden who served with the 45th Sherwood Foresters, a regiment which was garrisoned at Pietermaritzburg between 1843 and 1859. During Garden's time in Natal, between the late 1840s and early 1850s, he and Stanger became friendly. In 1853 Garden used Stanger's wardian case to send plants to Kew and the following year, when on visit to Britain, Garden took to Kew a wardian case filled with plants collected by Stanger,

Unfortunately for Kew the amicable partnership between Stanger and Garden was brought to a premature end by Stanger's much lamented death in March 1854. Garden retired from the army not long afterwards but later became a member of the Indian colonial service and sent Kew specimens from that subcontinent. When describing *Albuca gardenii* in *Curtis's Botanical Magazine* in 1855, Sir William Hooker acknowledged the sterling efforts made by this army officer.[50]

Though Stanger's death was deeply regretted by the colony and by Kew, the early 1850s had seen the arrival of two enthusiastic amateur botanists who were soon to make their mark on the development of Natal botany. The first to come, in March 1850, six months before McKen, was John Sanderson; the second was Dr P.C. Sutherland who arrived in November 1853.

John Sanderson was a domineering, sharp-tongued Scottish journalist. Years later Lieutenant Governor R.W. Keate remarked that Sanderson possessed 'the character of seldom agreeing with anybody about anything' and that he was 'a man of extremely bilious temperament'.[51] As a journalist on the *Natal Times,* and later as editor of the *Natal Colonist* and member of the legislative assembly, he fought many bitter battles. He was particularly hostile to the immigration of Indians to Natal from 1860 onwards.[52]

Yet there was another side to this colourful figure: Sanderson was a self-confessed dilettante in the study of natural history. He was interested in and knowledgeable about birds, animals and particularly plants, the last of which he collected and occasionally sketched. Kew's attention was initially drawn to him by a five-month collecting trip into the interior of southern Africa as far as the Magaliesberg mountains.[53] The most noted of the discoveries he made on this epic journey was found just a few miles from Durban at Field's Hill and again at Zwartkop outside Pietermaritzburg. This new genus of Liliaceae, named *Sandersonia aurantiaca* by Sir William Hooker, soon became popularly known as the Christmas bell.[54] For some 30 years Sanderson was to prove himself an excellent supplier of botanical specimens. Another notable example of Sanderson's endeavour for Kew was his despatch to the Hookers in 1862 of bulbs of the beautiful Natal paintbrush, *Haemanthus natalensis*.

---

49 *Stangeria paradoxa* is now referred to as *S. eriopus*.

50 *Curtis's Botanical Magazine*, (1855), t4842.

51 A.W. Bayer, 'Discovering the Natal flora', *Natalia*, 4, (December 1974), p.45; and Alan F. Hattersley, *The Natalians*, (Pietermaritzburg, 1940), p.143.

52 K.C.A.L.: folder 26944, M. Jowitt to Killie Campbell, 8 February 1949.

53 Rev. John Buchanan, *Revised list of the ferns of Natal*, (Durban, 1875), p.30.

54 *Curtis's Botanical Magazine*, (1853), t4716.

They flowered at Kew the following year.[55]

Sanderson's name is commemorated in many species including *Argyrolobium sandersonii*, *Basananthe sandersonii*, *Ceropegia sandersonii*, *Cissus sandersonii*, *Lissocrilus sandersonii* and *Senecio sandersonii*. Mention should also be made of the asclepiad which Sanderson discovered, *Caralluma lutea*. Sanderson's specimens and comments on plant illustrations proved invaluable to Harvey in his compilation of the first volume of his *Flora Capensis*, and Sir William Hooker was soon writing of 'our friend Mr Sanderson'.[56]

As well as despatching his own plants to Kew, Sanderson also sent to the Hookers plants from other collectors such as G.T. Fannin, John Meade and Robert Plant.[57] Sanderson was able to offer this service because of the position he held in Durban's horticultural society. In July 1851 he had been elected to the society's committee; by 1853 he was its secretary and in 1859 he became president of the society, a position he was to hold for over 20 years.

During this time he was generally of the opinion that the botanic gardens was mismanaged and he and McKen, in particular, had an uneasy relationship despite Sanderson's respect for McKen's knowledge of the local flora.[58] None the less Sanderson had the gardens' best interest at heart and frequently asked Kew for new species or for more specimens of old ones which had failed. As early as November 1852, when he had been on the committee for just over a year, he was asking Sir William Hooker for new tea plants for the gardens as those which McKen had brought out had all died. It is a reflection of the awareness of Natal botanists of developments overseas that Sanderson requested seeds of the giant *Victoria amazonica* (*V. regia*) at that time.[59] This was the same year that Richard Turner's waterlily house, purpose built to house the newly imported south American lily, was opened at Kew.

In June 1865 Sanderson wrote 11 pages of interesting commentary entitled 'Rough notes on the botany of Natal' as an appendix to J. Chapman's *Travels in the interior of South Africa*. Sanderson also included in this appendix 'Some observations on the botany of Natal', the text of a lecture delivered in Pietermaritzburg by Rev. Edward Armitage.

The second amateur doyen of Natal botany to arrive in those early years was another Scot, Dr Peter Cormac Sutherland.[60] He was less outspoken than Sanderson and was the most scholarly of Natal's botanical pioneers. As a young man Sutherland had led a rather nomadic life. He had been a sailor, venturing to west Africa in a guano ship and into the arctic regions on whalers. After qualifying as a medi-

Dr P.C. Sutherland (1822-1900)
This much-travelled Scot settled in Natal in 1853 and two years later became the colony's surveyor-general, a post he held for 34 years. He was an ardent plant collector, and was probably the most knowledgeable of the amateur plant-hunters of Natal. His office took him to many remote parts of the colony and its hinterland and he used these opportunities to collect for Kew. The beautiful Natal bottlebrush (*Greyia sutherlandii*) is named in his honour as is a Natal mountain and an island off Greenland.
(Courtesy: Natal Archival Depôt, photograph number C.116)

55 A. Batten and H. Bokelmann, *Wild flowers of the Eastern Cape Province*, (Cape Town, 1966), pp.22-3.

56 See *Curtis's Botanical Magazine*, (1855), t4850; *Flora Capensis*, I.9; and K.A.: Sanderson, 19 August 1859.

57 K.A.: Sanderson, 3 October 1858, 3 January 1859 and 4 July 1864.

58 See K.A.: Sanderson, 3 January 1859.

59 K.A.: Sanderson, 9 November 1852.

60 Britten and Boulger, 28, (1890), p.348; and Gunn and Codd, p.338.

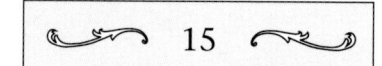

cal doctor he had acted as geologist to two expeditions (1850 and 1852) to north-eastern Canada in search of the ill-fated Captain Franklin. In 1852 he published a colourful two-volume account of the first expedition. The flora of Greenland had particularly attracted him and an island off Greenland's west coast was named Sutherland in his honour. But the scarcity of new plants in Greenland, combined with a fear for his health if he remained in the 'icy regions', led him to approach Sir William Hooker and request him to intercede with the colonial service for a medical appointment in the recently much publicised colony of Natal. It is interesting that Sutherland felt Hooker carried such sway with the colonial office and it is equally significant that Sutherland did not receive the desired posting.

Let down by Sir George Barrow of the colonial office, Sutherland, in consultation with Hooker, decided to emigrate to Natal on spec, which he did in the latter half of 1853. He was 31 years old and unemployed.[61] Four months later, following a not very successful interview with the lieutenant governor Sir Benjamin Pine, Sutherland wrote to Kew:

This is a truly charming country especially in a botanical point of view. Neither Sanderson nor McKen, both of whom I have met, will exhaust its extensive flora. They are very observing but their duties fix them to certain localities... If I could make a living here I should have much pleasure in examining the plants that may occur in the Materia Medica.[62]

Two weeks later Sutherland's fortunes changed and he was offered the post of government geologist at a salary of £200 per annum. In 1855, as a result of Dr Stanger's death the previous year, P.C. Sutherland was appointed Natal's surveyor-general, an office he held until 1887. It was an ideal position for a knowledgeable botanist. By the very nature of the job the surveyor-general had to travel throughout the colony laying out settlements, planning the routes of new roads, surveying private farms and crown land.

AS THE YEARS PASSED Dr Sutherland came to be regarded with considerable esteem in Natal, being active in colonial medicine, geology and politics; and from an early date he made his mark among fellow naturalists in the colony. Sanderson wrote to Joseph Hooker in April 1854, 'I am glad the colony has made the acquisition of such an enthusiastic follower of science'.[63] And to Robert Plant, Sutherland was simply 'a very estimable gentleman'.[64]

In these early years Sutherland corresponded with Kew. He supplied them with information on the colony's flora as well as with seeds and plants and occasionally dried specimens for the herbarium.[65] In 1857 he sent Kew 400 species of live plants collected in the vicinity of Durban. The previous year he had sent Kew heaths, grasses and a crop grown by Africans which he called 'ulsamin'. He remarked of this cereal:

Immediate steps could be taken to encourage the natives to grow it more largely for the purpose of exploitation as at present the natives produce nothing that can be exported except hides of cattle.[66]

In 1864 Sutherland became equally excited about a flax-like plant he discovered to the south of the colony in Pondoland. He expressed great hopes that this might prove a new and viable crop for Natal farmers. He dispatched specimens to Kew via Dr Charles Godwin and in due course Kew identified the plant as *Linum thunbergia*.

Best remembered for discovering 'one of the most remarkable of south

---

61 See Sutherland's letter to Kew from Port Elizabeth, K.A.: English letters, vol XXXIII, 18 October 1853.

62 K.A.: Sutherland, 11 March 1854.

63 K.A.: Sanderson, 8 April 1854.

64 K.A.; Plant, 15 April 1857.

65 *Kew Report,* (1864).

66 K.A.: Sutherland, 26 August 1856.

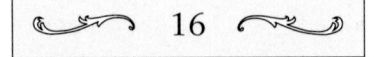

Hunting for animals and plants
Plant hunting was not just the preserve of impoverished botanists. The most influential of the colony often combined animal hunting with plant collecting. In this photograph from the early 1860s Lieutenant Governor Scott (with his foot on a chair) relaxes in camp with three of the Shepstone family. (Sir Theophilus is on the left). 'I am totally ignorant of botany', Scott wrote to Kew, but he discovered the Natal flame bush (*Alberta magna*) in a kranz in Karkloof. (Courtesy: Natal Archival Depôt, photograph number C.210)

east African shrubs', *Greyia sutherlandii*,[67] Sutherland is also commemorated in such species as *Begonia sutherlandii*, *Argyrolobium sutherlandii*, *Millettia sutherlandii* and *Vernonia sutherlandii*. Harvey in his *Thesaurus Capensis* (1859) made a special point of congratulating Sutherland for noticing *Erica alopecurus* in Natal:

It appears to have escaped the notice of all African collectors... and is interesting as being one of the few heaths that struggle into the sub-tropical regions North East of the colony.[68]

Besides these enthusiasts Kew was in contact with a number of people connected with the colony who had a general interest in botany. These included W. Bell, Arthur Clarence, the Natal postmaster general W.M. Collins, W.P. Robertson and William Stewart Mitchell D'Urban who sent Kew ferns from Natal in the early 1860s.

One of the most notable lieutenant governors of Natal, John Scott, was numbered among these amateurs. In 1856 at the age of 42 he came to rule the backwater colony of Natal: that his efforts to promote the new constitution and sagging economy during his eight-year term of office met with limited success was not through want of trying on his part. Scott's tenure is primarily remembered as the initial period of Indian emigration into the colony. But he was also active in other spheres, including the promotion of colonial agriculture. He was the first lieutenant governor to take a keen interest in the Durban botanic gardens, the only one in the colony at that time. Following a visit there in January 1862 he announced that the government would give the horticultural society an annual grant of £350 in return for nominated representation on its council and a promise that the institution would do more to promote the interests of the colony.[69]

Scott was fortunate in having the services and advice of Sutherland. The latter had selected specimens of

67 *Flora Capensis*, (1860), I.9-10.
68 W.H. Harvey, *Thesaurus Capensis*, (Dublin, 1859), I.31.
69 D.L.H.M.: Horticultural society minute book, H1130B, 21 January 1862.

Bishop John William Colenso, (1814-83), surrounded by books and scientific apparatus in his study at Bishopstowe outside Pietermaritzburg. Sir Joseph Hooker admired Colenso's stand against religious fundamentalism yet Sir Joseph noted to Charles Darwin that the bishop was 'sanguine and unsafe'. His father, Sir William Hooker, corresponded with the bishop. In one consignment he sent Colenso four varieties of azalea, a shrub now synonymous with Pietermaritzburg. (Courtesy: Killie Campbell Africana Library)

timber from the Port St Johns region in 1860, which it had been hoped might have been used in harbour construction at Durban. The colonial services, however, decided to use Baltic timber should the work be undertaken. Sutherland had told Kew of the consignment of indigenous timber en route for testing in Britain and Sir William Hooker promptly wrote to Scott, who was then temporarily in London, expressing the hope that Kew might eventually have the samples. Sutherland's specimens of timber from the banks of the Mzimvubu were lying at the London docks; Scott duly presented them to the Kew museum and offered, 'I am, I regret to say, totally ignorant of botany but I should none the less be glad to be of any use in procuring you specimens of Natal produce'.[70] Shortly after this Scott and Sir William Hooker met at Kew. It is a reflection of the esteem that Sir William had attained in the public mind that the lieutenant governor of Natal approached Kew under the auspices of a letter of introduction from 'my old friend Mr Conyngham of the Foreign Office'. None the less the warmth of subsequent correspondence indicates that John Scott and Sir William got on well together.

Three years later Scott had a minor success when he discovered a flowering tree growing out of a precipice at Karkloof. It was unknown to Sutherland so Scott sent a dried specimen in his despatch bag to Kew where it was identified as *Alberta magna*, now popularly known as the Natal flame bush. Kew regarded this as a rare South African plant and asked Scott to procure some of its seeds for them.

The following year, 1864, Scott left the colony. During his tenure of office in Natal, he placed the seal of official approval on Natal's support of Kew and set a pattern of senior government officials being interested in the colony's flora. At times this interest appeared to be to the detriment of what some regarded as more pressing matters.

70 K.A.: Scott, 24 October 1860.

Another amateur who had no great knowledge of botanical matters, yet was a celebrity in his own right, was the controversial anglican bishop John William Colenso. He corresponded from time to time with Sir William Hooker on behalf of himself and Miss Baker. The botanical worth of what Colenso sent to Kew was not great. Apart from unnamed bulbs and seeds, the bishop's contributions were listed by Kew as *Strelitzia, Hibiscus, Crinum* and *Gloriosa*. Conversely, however, Kew met Colenso's request for 'a few choice English seeds or plants'.[71] These were despatched in a wardian case via Liverpool to Natal on 1 March 1855. The case was listed as containing the following flowers and woody plants: four varieties of *Azalea*; four varieties of *Dahlia*; four varieties of *Camellia*; two varieties of cedar; three varieties of cypress; two varieties of oak; three varieties of tea; *Gardenia*; Jamaican ginger; China grass; monkey puzzle tree; Moroccan argania tree; osage orange tree; and walnut tree; as well as a *Hebeclinium, Escallonia macrantha, Brunsfelsia calycina, B.hydrangeaeformis, Rondeletia speciosa, Cereus macdonaldiae* and *Amomum cardomomum*. In addition Kew sent Colenso 200 packets of flower seeds.

Frances Colenso, the bishop's wife, was an amateur botanical artist. In 1865 she wrote to Mrs Katherine Lyell, a mutual friend of the Colensos and Hookers:

We have been here [the mission station, Ekukanyeni] some weeks and are not yet in order, which is partly the fault of Dr Hooker who supplied me with 500 kinds of flower seed. To sow these, beds had to be prepared.[72]

THE INTRODUCTION TO THE colony via Colenso of many exotic garden flowers was an important horticultural development, but Bishop Colenso was essentially significant in the Kew-Natal link because his rejection of the fundamentalist interpretation of the Old Testament in favour of rational explanation, his advocacy of secular representation on church councils and his compassionate approach towards the Zulus led not only to a schism in the anglican church in southern Africa but also to a public outcry in Britain and Natal. For Joseph Hooker the Colenso episode was part and parcel of the new age of rational enquiry, spearheaded by his friend Charles Darwin. Colenso had visited the progressive scientific 'X club'[73] of which Joseph Hooker was a founder member. The two men had also met at Kew and Joseph Hooker had stayed with Colenso when the latter was in Britain. Hooker was much taken with the bishop's 'calmness, his dignity and charity towards his opponents', but he was strongly opposed to Colenso remaining in Natal. On the 16 February 1864, shortly after Colenso was convicted of heresy,[74] Joseph Hooker wrote to Darwin:

His holding his bishopric in Natal can only breed intolerable confusion and do his cause mischief; and as to his going out to convert Zulus, why he has Christians here to convert, and the Zulus are not worth a thought...I have seen a good deal of him, and consider him sanguine and unsafe.[75]

Hooker subscribed anonymously to the fund for Colenso's appeal. This episode had brought the colony of Natal to the attention of the younger Hooker, Joseph, at a time when he was assistant director and just about to take over directorship following his father's death in 1865. Thus the new regime at Kew was as aware of Natal as had been the old.

71 K.A.: Colenso, 1 August 1853.

72 Wyn Rees, (ed), *Colenso letters from Natal*, (Pietermaritzburg, 1958), pp.98-9, 240 and 421.

73 The wives of the X club members were referred to as 'Y Vs'.

74 The privy council overruled this decision in 1865.

75 Huxley, II.58-9. For Hooker's friendship with Darwin, see W.B. Turrill, *Joseph Dalton Hooker: Botanist, explorer and administrator*, (London, 1963), pp.74-98.

Chapter 2

Early plant exchange

THE MAIN PURPOSE of Kew's interest in colonial Natal was the acquisition of as many species of its flora as possible, whether in the form of live plants, roots, bulbs, seeds, pressed and dried specimens or the occasional exhibit for one of Kew's three museums. The *quid pro quo* was the despatch to Natal of plants or seeds which might either be economically viable or simply attractive in settlers' gardens, possibly reminding them of 'Home'. Records of this exchange are to be found in the inward and outward exchange books and the plant list volumes which are housed in the Kew archive.

Much of the correspondence between Kew and Natal concerned the numerous problems encountered in this exchange. Collecting in the colony throughout the Victorian era was always fraught with difficulties. The humid subtropical climate along the coast made collecting difficult, especially in the wet summer months. Indeed the deterioration of specimens over the duration of a collecting trip was a problem throughout the period. Several solutions to this were devised but these usually meant killing the plant. Sanderson found that ferns could be preserved if plunged into hot water.[1] More common was the use of a form of wooden slatted press, in which plants were pressed and dried for preservation. But the main problem was that most collectors were unfamiliar with techniques of preserving specimens. In 1858 Kew sent the horticultural society in Durban instructions on how to preserve plants but this involved, at least in the case of succulents, the frequent turning of specimens for several weeks which was hardly practical while out collecting in the veld.[2] A year later Harvey's *Thesaurus Capensis* contained instructions for drying, but not every collector possessed this volume. In the early years in the colony McKen was using the 1823 edition of Thunberg's *Flora Capensis*, which had a Latin text. Natalians writing to Kew constantly complained that they had neither a good field manual nor practical instructions on plant preservation.[3]

Kew did try when possible to provide collectors with equipment. Robert Plant is recorded as having asked for a tent to use when out collecting and Kew sent Gerrard a portfolio in which to carry his specimens.[4]

Equally problematic were the logistics of travelling in the colony. Even on horseback, which was the quickest form of transport into the interior, the 50 mile (80 kilometre) ride from Pietermaritzburg to Durban took the best part of a day. The terrain was often difficult to traverse; roads were few and road-building spasmodic. Where roads did not exist a

---

1 K.A.: Sanderson, 6 February 1865.

2 K.A.: Sanderson, 31 October 1858.

3 For example, see K.A.: Sutherland, 4 June 1854; and Colenso, 1 August 1854.

4 K.A.: Plant, 16 January 1850; and Gerrard, 16 March 1865.

horse could stumble and fall, as happened to John Sanderson who broke his collarbone in this way when out collecting. Furthermore, as Dr Sutherland discovered, there was a limit to the amount of specimens a horse with rider could carry.[5]

The ox wagon was generally regarded as the ideal vehicle for plant collecting and it remained so throughout the Victorian era, though later it was used in conjunction with the railway. The principal advantage of the ox wagon was its roominess so that large plants, such as palms and cycads could be transported undamaged and with some root soil. The collector could also carry presses, spades and foodstuffs, such as meal and coffee. The slow-moving and sturdy ox wagon was also perfect for gradually traversing hard but interesting countryside that was rich in plant life. In crossing rivers, labouring along the floor of kranzes or trundling over the dry and rugged veld, the vehicle was, in the collector's eye, far superior to any horse-drawn cart.

But the ox wagon had its disadvantages. Travel was indeed slow even where roads existed: the journey from Pietermaritzburg to Durban could take three or four days. If one ventured into Zululand there was a very real danger of the oxen either being attacked by wild animals or, more frequently, falling sick and dying. Robert Plant in his famous first Zululand expedition encountered such hazards. Yet another problem about the ox wagon was cost: in the 1850s a wagon and span of oxen could be bought for about £150 or hired for 15 shillings a day.

Leaving aside the difficulties of transport, the remote areas penetrated by these plant collectors held many other dangers. Though African peoples were usually friendly to collectors and often assisted them, this could not be guaranteed in time of war: for instance, in the early 1850s, as we have seen, Plant stumbled on a war in Zululand. The threat posed by lion, buffalo, hippopotamus and snakes made it necessary for the plant collector to carry a gun on lengthy trips. Yet even near Durban collectors could encounter danger. At Sea Cow Lake, just north of the town, McKen was chased by a hippopotamus. He managed to escape by climbing a tree, where he had to wait for four hours before the animal moved away. When collecting in the Seychelles Robert Plant nearly fell off the edge of a precipice and noted of the accident, 'I had not been nigher a catastrophe since I stood across the alligator at Sea Cow Lake'.[6] Ticks were a constant irritation for all collectors, and for those who ventured into the 'fever-ridden swamps', with their often dramatic and colourful flora, there was the serious threat of disease: at best there

HAEMANTHUS INSIGNIS (Blood Flower). This member of the family Amaryllidaceae is now known as *H.magnificus*. A bulbous perennial, it produces only one flower a season. It first flowered in Kew in a cool frame in August 1853. It is one of the most attractive of Natal wild flowers.
(*Curtis's Botanical Magazine*, T4745 [1853])

5 K.A.: Sutherland, 18 March 1864.

6 See Daphne Child, *A merchant family in early Natal: Diaries and letters of Joseph and Marianne Churchill*, (Cape Town, 1979), p.24; and K.C.A.L.: KCM 1867, Notebook of Robert Plant, 1853-1854.

A wardian case The invention of the wardian case in the 1830s revolutionised the transport of plants over long distances. Firmly bedded in a wardian case, plants protected from the elements, had a reasonable chance of survival when sent to Kew from Natal or other distant parts of the globe. Most wardian cases exchanged between Kew and Natal were less ornate than this illustration.

7 Cape government notice no.387 (1857). See also Cape government circular no.78 (10 February 1858).

8 K.A.: Sanderson, 31 October 1858.

9 K.A.: Sanderson, 12 August 1859.

10 For example, K.A.: Sanderson, 6 February 1865.

11 Huxley, I.8

were mild attacks of malaria, at worst, death.

Under the provisions of a Cape government notice issued in 1857 any specimen of flora, if addressed to the colonial secretary, might be sent free of charge to Cape Town through the Cape's internal postal service.[7] In Natal this did not apply though later the Natal Government Railway guaranteed to transport free any botanical items for the Durban or Pietermaritzburg botanic gardens. But as the railhead had by 1880 reached only Pietermaritzburg this was not to help until towards the end of the century.

THE PROBLEMS OF PLANT collecting in the colony's early years were counterbalanced in the eyes of the collectors by the adventurous nature of the enterprise and by the possible rewards it offered. The spirit of these men is well caught in a letter written by Sanderson to Kew in October 1858. Collecting near the coast north of Durban, he and a friend found:

a beautiful shrub growing among the stones in the bed of a stream. There it was, one mass of matt delicate powder pale lilac campanulate flowers, and we shouted and shouted again with delight.[8]

The problems of collecting plants in the colony were multiplied when it came to transporting the specimens overseas to Kew. Few were as unfortunate as John Sanderson who, during a voyage to Britain in 1859, discovered when his ship sailed out of Cape Town that his precious collection of plants in their wardian case had been left on the dockside.[9]

Plants were conveyed to Kew by entrusting them to one of four agents: a traveller, a government official, a shipping company or the postal service. In theory the cost of sending a collection was prohibitive. In 1854 an ordinary package sent to Britain cost between 10 shillings and £2. Invariably, however, specimens travelled free of charge.[10] Either they were taken with a traveller's own luggage, or by the admiralty or colonial service, or even occasionally by a steamship company.[11] The problem for Kew if specimens travelled by these irregular means was that they had no bills of lading, which necessitated a Kew official going to the docks to find and identify the valuable cargo.

By the time live plants were being transported between Natal and Kew the glazed or wardian case was the principal type of container. The sturdy glass-sided case, which could vary greatly in size, was designed in the 1830s by Dr Nathaniel Ward. Plants rooted in dampened soil in the base of the case, or in pots, were protected from the hostile, salty-aired shipboard environment yet allowed light through the glass. If properly sealed the plants were able to create their own microclimate inside the case. Such cases were originally designed to protect plants from the polluted atmosphere of industrial British cities and so could sometimes be extremely ornamental; those used on the long-haul sea voyages were more utilitarian. The invention of the wardian case revolutionised the world's agricultural economy: for the first time large

ON THE CONVEYANCE OF PLANTS AND SEEDS ON SHIP-BOARD.

quantities of plantation crop plants could be transported over thousands of miles of ocean. Robert Fortune importing 20 000 tea plants to the Himalayan region from Shanghai is perhaps the most famous illustration of this but the movement of species of chinchona, rubber, coffee and sugar cane to various parts of the tropics and subtropics was equally important.

As a generalisation Joseph Hooker's statement to Harry Bolus of the Cape that, 'all steamships hate Ward's cases' was accurate. McKen was fortunate when Byrne's emigration company allowed him to keep his wardian case on the protected quarter-deck during the voyage to Natal in 1850.[12] Bishop Colenso obtained a similar concession from his ship's captain in 1854.[13] But too often the wardian case was roughly man-handled into the hold where there was no sunlight and where the glass was easily broken. If the latter occurred, the plants were exposed to the salt-laden air as well as to attack from rats and insects. Equally disastrous was overwatering of the plants or, as was more likely on sailing ships where fresh water was at a premium, underwatering.

Yet another problem of plant exchange was that when plying between Britain and Natal the ships had to pass through the hot equatorial region. The longer this took, the more danger the plants were in: in the early 1850s a sailing ship might be expected to take 60 days to reach Cape Town from Britain and then a further two weeks to make Durban. Thus the plants had to survive 11 weeks, and often longer, under harsh conditions. The steam mail ships, on the other hand, could reduce this time to seven or eight weeks.

The Natal-Kew correspondence and occasionally the local colonial press contain reference to the problem of plant mortality in transit. In August 1854 the *Natal Mercury* noted that five-sixths of the 30 plants which Kew had despatched to Natal in a wardian case had arrived dead.[14] Of an earlier consignment Sanderson reported to Kew, 'The vessel unfortunately had a passage of some four months, and when opened the plants appeared to have been killed by damp'.[15] Similarly in 1855 seven out of 19 specimens which Robert Plant sent to Kew died on the voyage,[16] and nearly a decade later Kew received seeds from Dr Sutherland 'very much eaten by beetles'.[17] A large wardian case was despatched by McKen in 1862 to John Smith, the curator of Kew: the plants were placed in pots and the spaces filled with palm and fruit tree seeds, 'all natives of Natal'. This was a common practice if pots were placed in the case; Dr Sutherland is recorded as having filled the space between the pots with African-produced tobacco leaves. In the instance of McKen's case, Kew noted that the plants had arrived in good condition but that 'only four or five seeds were found and nearly all those were dead'.[18]

Despite losses many wardian cases successfully conveyed their plants safely across the 7 000 miles (11 000 kilometres) which separated Natal and Kew. Nine or ten months seems to have been the average time it took a case to return from Natal on a round trip, but it was often longer. One case sent to McKen in May 1853 did not arrive back at Kew until May 1855 and occasionally 'not returned' is noted beside the record of a case despatched to Natal.[19] Very often the curator of the Durban botanic gardens had to repair the container before it could be returned. A case might also be delayed in the gardens, waiting for different collectors to bring specimens to be packed into it, or for a forthcoming overseas trip by some unfortunate

12 K.A.: South American letters, vol LXX, McKen (London), 24 June 1850.
13 K.A.: Colenso, 1 August 1854.
14 *Natal Mercury*, 2 August 1854.
15 K.A.: Sanderson, 8 April 1854.
16 K.A.: Exchange book, inward, (1848-1859), 19 May 1855.
17 K.A.: Exchange book, inward, (1859-1867), 20 July 1864.
18 *ibid*, 30 August 1862.
19 These are possibly clerical errors; see K.A.: Exchange book, outward, (1848-1859), 22 January 1853 and 1 March 1855.

THUNBERGIA NATALENSIS A Mr Cuming sent this species to the famous nursery firm of Messrs Veitch in England, where it first flowered in July 1858. It is a member of the Acanthaceae family, which likes secluded damp places. (*Curtis's Botanical Magazine,* T5082, [1858])

who had been requested to accompany and care for the precious load.

What exactly was exchanged between Natal and Kew in those early years? The inward exchange (to Kew) is much more difficult to analyse than the outward because the collectors were either inexperienced in identifying the rarer species or simply quite ignorant of a plant or seed's identity. The inward exchange volumes contain such comments as (sic): 'Lycopodium like L. flagella'; 'New genus with red flower like Zephyranthes'; Black with red fringed ocilla'; 'Blue seeded leguminosae shrub like Viborgia'; and 'Convolvulaceae seed with woolly tomentum'.

The annual volume of plants and seeds sent to Kew during the 1850s and early 1860s was not as great as it was to become during the rest of the nineteenth century, but in some respects it set a pattern. Any cycad or *Strelitzia* was sent because these species were very popular with European botanists and plant lovers. Some orchids, ferns and succulents were sent but not in such numbers as later. The early shipments of plants to Kew from Natal more often comprised certain species, *Crinum, Clerodendron, Gladiolus* and jasmine. Strangely no aloes seem to have been sent in these early years. On 30 August 1862 Kew received a wardian case from McKen, the contents list which is worth giving because of the variety of identified plants it contained:

1. Rubiaceae & Ternstroemia
2. *Clerodendrum myricoides*
3. *Trichilia* sp & a small bulb from swamp, Umbilo
4. *Jasminum* sp
5. *Dombeya burgessiae*
6. *Rhus* sp
7. *Halimum* sp & *Cissus*
8. *Oxyanthus natalensis*
9. *Myurus inaequalis* (?)
10. Ochnaceae sp, *Xanthoxylum* sp & *Carissa*
11. *Stapelia* sp.: from Umvelos River, Zulu Country

(No number 12 is given)

13. *Guttaria caffra* (?) & *Royena* sp
14. *Cissus fragilis*
15. *Euphorbia* sp: grows a large 30-40ft high
16. Asclepiadaceae sp: strong growing climber like *Hoya*
17. *Rhus* sp[20]

Samuel Hereman's second edition of *Paxton's Botanical Dictionary,* published in 1868, gives only 13 new species of Natal plants introduced into Britain. Though this list is grossly incomplete, as later editions of the *Flora Capensis* were to prove, it is interesting to note the names of these species as their inclusion in the volume denotes their being familiar in Britain by the mid-1860s. The 13 species are: *Sandersonia aurantiaca* (introduced 1851); *Stangeria paradoxa* (1855); *Thunbergia natalensis* (1857); *Richardia hastata* (1858); *Methonica plantii* (*Gloriosa plantii*) (1859); *Haemanthus natalensis*

20 K.A.: Exchange book, inward, (1859-1867), 30 August 1862.

24

(1862); *Hypoxis elata* (1863); *Webbia pinifolia* (1863); *Kleinia fulgens* (1865); *Begonia geranioides* (1866); *B.sutherlandii* (1867); *Ipomoea gerrardii* (1867); and *Talbotia elegans* (1867).

The nature of the Natal climate meant that plants sent to Kew usually had to be kept under glass. Increased room for such delicate flora was provided by the erection of a cactus and succulent house in 1855 and of the central block of the temperate house in 1862. The latter in particular supplemented the great palm house. The acquisition of the Bentham and Blomfield libraries and herbaria in the mid-1850s, a decade before those of Sir William Hooker were purchased, encouraged Kew in building up their specimens of dried plants. Though not in such voluminous quantities as later in the century, herbarium material was already being despatched in this early period from Natal to Kew.[21] In addition the opening in 1857 of Kew museum I, opposite the great palm house, and the conversion of the orangery to museum III[22] five years

HYPOXIS LATIFOLIA *Paxton's Botanical Dictionary* rather unfairly said that this was a genus of no great beauty. This species was brought to Kew from Natal by Captain Garden in the European spring of 1854. (*Curtis's Botanical Magazine*, T4817, [1854])

The interior of the Orangery, then being used as the wood museum
Wood specimens sent from the southern border of Natal collected in 1860 by the colony's surveyor general, Dr P.C.Sutherland, were exhibited in this museum.
(From, *The Gardeners' Chronicle*, 4 November 1876).

21 *Kew Report*, (1864).
22 See *The Orangery: Royal Botanic Gardens, Kew*, (H.M.S.O., c.1971).

25

later encouraged Natal collectors to send to Kew the occasional 'oddity' Captain Garden was particularly helpful in this respect. Sanderson was also most enthusiastic. In December 1857 Sanderson wrote to Kew, sending:

two or three trifles for you in the way of Kafir curiosities for your museum, though it will, I fear, be difficult to give you the names of the plants they are made from.[23]

In 1859 he sent 'a couple of skulls of natives, Kafir and Hottentot'.[24] Human skulls were not the only strange offerings to be sent by Sanderson: in 1854 in a letter which accompanied a wardian case for Kew, he noted:

In this case I have put four tree frogs which I hope may arrive safe with you. Were it summer I could have sent you two or three species of them as well as chameleons but at this season they are scarce.[25]

The outward exchange from Kew to Natal consisted of three types of plants: those suitable for economic exploitation; those ornamental trees and shrubs and flowers familiar to colonists who were keen to create English-type gardens; and, to a much smaller extent in the early period, trees of value for their timber or bark.

As we have seen, McKen was entrusted with a large collection of plants of possible economic worth to the colony when he emigrated from Britain to Natal in 1851. The wardian case which Sir William Hooker presented to him in July of that year contained some 29 species, including allspice, arrowroot, breadfruit, camphor, chocolate nut, chinchona, cinnamon, cocoa plum, Arabian coffee, ginger, guava, jackfruit, litchi, mango, pawpaw, black pepper, red cassia, Assam and Paraguay tea and wompee as well as five ornamental shrubs: *Carapa guianensis, Cephaelis ipecacuanha, Jambosa vulgaris, Roupellia grata* and *Xanthochymus pictorius*. Sam Beningfield, a prominent settler and active member of the horticultural society, was sent by Kew in 1859 a wardian case which contained among other plants four varieties of pineapple, listed as queen, marcone queen, prickly cayenne and smooth. Cork oak (*Quercus suber*) was sometimes despatched to Natal by Kew.[26]

In addition to the garden plants listed in the previous chapter which were sent by Sir William Hooker to Bishop Colenso, in 1853 Kew sent Dr Stanger five varieties of *Chrysanthemum*, four of roses and dahlias, three of camellias and geraniums and one variety each of *Rhododendron* and *Magnolia*, as well as a vine, a cork oak and raspberry and fig plants.[27] In the same year McKen was sent, among other plants, specimens of cedar and cypress and a monkey puzzle tree.[28]

---

23 K.A.: Sanderson, 8 August 1857.

24 K.A.: Sanderson, 12 August 1859.

25 K.A.: Sanderson, 20 July 1854.

26 For example, K.A.: Exchange book, outward, (1848-1859), 22 January 1853 and 19 May 1853.

27 *ibid*, 22 January 1853.

28 *ibid*, 19 May 1853.

Chapter 3

Imperial Kew and the heyday of Natal botany (1866-1883)

Kew in the new era

THE MID-1860s PROVED to be a turning point for Kew and for Natal's relations with Kew. For both a period of pioneering growth came to a rather sudden end. On 12 August 1865 Sir William Hooker died aged 80, having recently completed his magnum opus, the five-volume *Species Filicum*. Active until shortly before his death, Sir William had been the mainstay of the small Kew establishment for a quarter of a century. He rescued the gardens from obscurity and neglect and laid the foundation of their greatness as a botanical and imperial institution. Though of considerable ability in the scientific field, his achievements were in no small part the product of his tact, diplomacy and pleasant gentlemanly manner.[1] The year 1865 also witnessed the death of the collector William Gerrard; and on 15 May 1866 his patron, William Harvey, died at the home of Sir William Hooker's widow in Torquay. In a letter to Joseph Hooker, McKen simply stated, 'We all miss Dr Harvey very much.'[2]

The end of the old era also saw the retirement in 1864 of 'old Jock', John Smith, who had served Kew as curator for 42 years and had witnessed the gardens' decline after Sir Joseph Banks' death, its stay of execution thanks to Dr John Lindley and finally its rebirth under Hooker. Though in later years he was somewhat cantankerous and nearly blind, Smith's retirement left a serious gap, with Joseph Hooker admitting that there was now no one at Kew capable of naming ferns.[3]

The accession to the Kew directorship of Dr Joseph Dalton Hooker (1817-1911), following his father's death, was by far the most momentous occurrence in Kew's history in the mid-Victorian era. Joseph Hooker had visited Cape Town as a young man when he accompanied Ross's famous expedition to the antarctic and the southern seas. Hooker did not think much of Cape Town though he liked the old Dutch East India Company garden in the centre of the town.[4]

In the late 1840s Hooker was collecting in northern India, an expedition which led to the introduction of 28 new species of *Rhododendron* to Europe.[5] On returning to Britain, Hooker worked on the scientific side at Kew, supported financially by his father. Finally, in 1855, William Hooker persuaded the government to appoint his son as assistant director of Kew. By the time he succeeded his father a decade later Joseph Hooker had made a name for himself as an outstanding

---

1 Bean, p.49; and William Gardener, 'The Royal Botanic Gardens at Kew: The work of Sir William Hooker and John Lindley', *History Today*, XV, 10, October), pp.688-96.

2 K.A.: McKen, 10 February 1867.

3 Huxley, II.10; Ronald King, *The World of Kew*, (London, 1976), p.52; and *A short account of the fern collections*, Royal Botanic Gardens, Kew, (Kew leaflet, 1976).

4 The old Dutch East India Company garden became the Cape Town botanic gardens in 1849.

5 See J.D. Hooker, *Himalayan journals*, (London, 2nd edition, 1855).

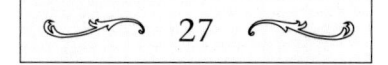

botanist and administrator, indeed, as Sanderson wrote to him, 'It seems quite natural that you should succeed... as director'.[6]

Less tolerant of inefficiency than his father, Joseph had none the less much of the old man's charm and generosity. Not only did Joseph Hooker engross himself in scientific study and debate in Britain but he also developed a great enthusiasm for exotic vegetation. Even when director he occasionally managed to escape the burden of running Kew to venture abroad. In 1871 he visited north Africa and six years later north America, 'to taste the delights of savagery again'.[7] Not surprisingly Hooker's major interest lay in British colonies such as Natal, where he could correspond with and receive specimens from British settlers. This was of particular value to him in his study of transoceanic vegetation links, such as the relationship between the flora of southern Africa and that of the south Atlantic islands.[8]

This enthusiasm for colonial flora was not shared by Joseph Hooker's superior, A.S. Ayrton, the chief commissioner of works in Gladstone's first liberal government and a quarrelsome individual. The department of works had assumed responsibility for Kew in 1851 and so far had not greatly interfered with the gardens. Encouraged by the malicious zoologist Professor Richard Owen, who had accused the Kew herbarium of achieving nothing more than 'attaching binomials to dried foreign weeds',[9] Ayrton issued a reprimand to Hooker over the manner in which Kew was run: the first such reprimand since before Sir William Hooker's appointment in 1841. Soon Ayrton found himself arrayed against Darwin, Huxley, Lyell and most of the other great British scientists of the age. The assault on Hooker's management of Kew lasted for two years and resulted in Ayrton's transfer to another government department in 1872 and to the eventual demise of his political career.

Through the public enquiry and parliamentary debates, Joseph Hooker, who was greatly upset by the crisis, emerged as something of a public hero at home and in the colonies. In Natal botanists were beside themselves with rage at how Hooker had been treated. When the storm erupted John Sanderson told Hooker bluntly, 'Mr Gladstone's conduct seems to me very contemptible'.[10] Dr Sutherland, having expressed his delight to Hooker that Ayrton 'has been brought to his knees', in April 1873 wrote, 'I am not sorry Mr Ayrton has gone to the wall, the more extraordinary commissioner of works there can hardly be than he was'.[11]

The repercussions for Kew of the Ayrton affair were considerable. In 1874 Gladstone, who had remained fairly neutral on the subject of Kew, was replaced as premier by the imperialist conservative prime minister Disraeli. A keen tree planter himself,[12] Disraeli looked with sympathy on Kew and certainly had no objection to 'dried colonial weeds'. As an official indication of the government's approval of his work, Joseph Hooker was offered a K.C.M.G. for services to the colonies. Hooker felt that it would be improper for him to accept membership of this order of 60 knights as he felt it was intended as a colonial institution, so he declined the offer. In 1877, however, he was created knight commander of the star of India.[13] The scientific world had already honoured him in 1873, by making him president of the Royal Society.

Disraeli also allowed himself to be persuaded by Hooker into granting a pension to the Kew botanical artist Walter Fitch after being shown Fitch's illustration of the great Victoria lily. The Disraeli government sanctioned

---

6 K.A.: Sanderson, 8 March 1866.

7 Joseph Hooker to Charles Darwin, quoted in Allan, p.219.

8 Huxley, II.234 and 244.

9 Allan, p.222; Wilfred Blunt, *In for a penny: A prospect of Kew Gardens,* (London, 1978), p.162; and Huxley, II.174.

10 K.A.: Sanderson, 1 July 1872.

11 K.A.: Sutherland, 24 February 1873 and 23 April 1873.

12 G.E. Buckle, *The life of Benjamin Disraeli,* (London, 1920), VI.638.

13 Huxley, II.149.

Sir Joseph Dalton Hooker (1817-1911) Sir Joseph was the son of Sir William Hooker. He was assistant director of Kew from 1855 until he succeeded his father as director on the latter's death in 1865. As enthusiastic a botanist as his father, Sir Joseph presided over Kew until his retirement in 1885, a period during which the institution was involved in the staffing of many new colonial botanic gardens. William Keit's appointment to the Durban botanic gardens was arranged by Sir Joseph in cooperation with Dr David Moore of Glasnevin.
(Courtesy of the Royal Botanic Gardens, Kew)

the recreation of the post of assistant director of Kew, a position which had been abolished 10 years earlier in 1865. The stern but able William Thiselton-Dyer was appointed to this post. Two years later, in 1877, Thiselton-Dyer married Joseph Hooker's daughter, thus ensuring the continuation of the Hooker dynasty at Kew.

Increasingly Thiselton-Dyer dealt with the day-to-day colonial correspondence, especially that connected with the herbarium. This department had been under Alan A. Black until 1863. From then until 1890 it was presided over by Joseph Hooker's lieutenant, the impressive and scholarly Professor Daniel Oliver, who combined the position with being professor of botany at London University. Oliver had under him, among others, N.E. Brown and John Baker who later succeeded him.

In 1877, a year after the completion of the Kew Jodrell laboratory for the study of plant physiology and

'River scene Victoria lilies', by Walter Hood Fitch. Plate T4275 in *Curtis's Botanical Magazine,* 1847. It is said that later when Sir Joseph Hooker showed the prime minister, Benjamin Disraeli, the plates of the giant Victoria lily done by Fitch. The great imperial statesman was so impressed by their majesty that he granted Fitch a pension.
(From, *Curtis's,* supplied by the National Botanical Institute, Pretoria)

Professor Daniel Oliver (1830-1916) Though being described as being a mixture of 'extreme modesty and retirement', Professor Oliver had an influential voice in Victorian Kew. He worked there from 1858. From 1864 to 1890 he was keeper of the herbarium and library. He was especially interested in Africa's flora and was often called upon to identify Natal plants sent to Kew. He edited the first three volumes of the flora of tropical Africa. Between 1861 and 1888 he held the chair of botany at the University of London.
(From, *Kew Guild,* 1898)

anatomy, a new wing of the herbarium was opened. The expansion of the scientific side of Kew, both in staff and in buildings, made it much more of an institution as distinct from William Hooker's and George Bentham's personal establishment.

The purchase of Sir William's private library and herbarium for Kew in 1867 marks the beginning of this institutionalisation. The result was dramatic. There were additions to the gardens themselves, most notably the erection at the end of the 1860s of the 'T range' of heated plant houses.[14] The new curator, also named John Smith like his predecessor, after some hard work had restored to their former glory the greenhouses which had been somewhat neglected by John Smith I.

The new curator also had to contend with devastating frost and heavy snow in January 1867.[15] Though he continued improving the layout of the gardens, by the 1870s its pioneering stage was fast drawing to a close. The major innovative developments now lay on the scientific side of Kew's operation. It is well illustrated by the large number of volumes published by or in association with Kew in this period. These include *Curtis's Botanical Magazine, Genera Plantarum, Icones Plantarum* and the annual *Kew Report.* In addition under Joseph Hooker's general direction the colonial flora project was pursued with vigour. Between 1866 and 1883 the New Zealand and Mauritius and Seychelles floras were completed and the British India and tropical Africa floras com-

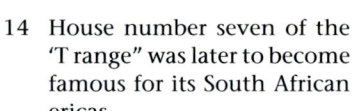

---

14 House number seven of the 'T range" was later to become famous for its South African ericas.

15 Allan, p.217

CURTIS'S
BOTANICAL MAGAZINE,
COMPRISING THE
Plants of the Royal Gardens of Kew,
AND
OF OTHER BOTANICAL ESTABLISHMENTS IN GREAT BRITAIN,
WITH SUITABLE DESCRIPTIONS;
BY
SIR WILLIAM JACKSON HOOKER, K.H., D.C.L. Oxon.,
LL.D., F.R.S. and L.S., Vice-President of the Linnean Society, and Director of the Royal Gardens of Kew.

VOL. IX.
OF THE THIRD SERIES;
(Or Vol. LXXIX. of the Whole Work.)

"Another Flora here, of bolder hues,
And richer sweets."

LONDON:
LOVELL REEVE, HENRIETTA STREET, COVENT GARDEN.
1853.

The frontispiece of *Curtis's Botanical Magazine* for the year 1853. It was in this Kew-based journal that many Natal plants were first described; three Natal plants are contained in this volume alone.

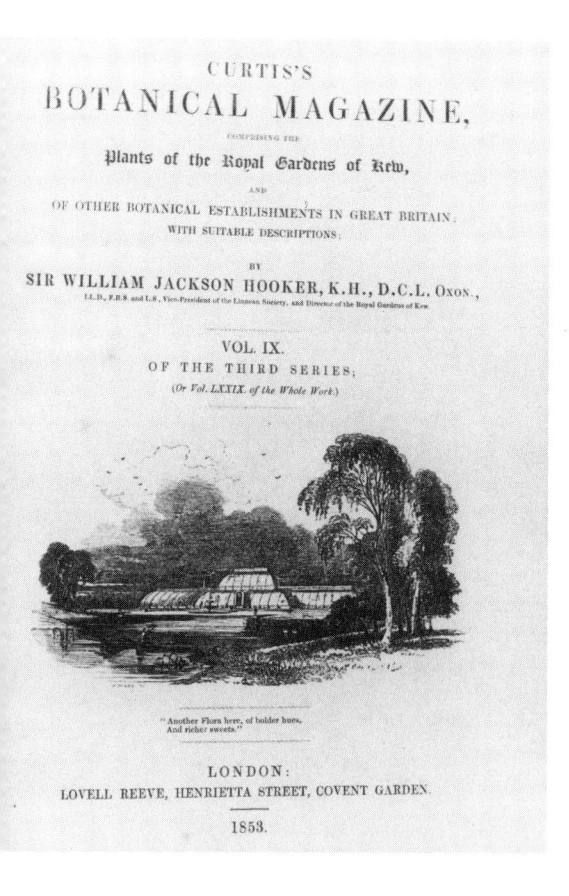

menced. The death of Harvey had led to the shelving of the *Flora Capensis* but in 1868 Joseph Hooker edited a second updated edition of Harvey's *Genera of South African Plants*.

Natal and the new era

THE DEVELOPMENT OF KEW into the world's principal botanical institution with an annual budget of £12000 was the consequence not only of the high calibre of the men working in the Kew establishment but also of the popular enthusiasm for plants. By the mid-1860s the Victorian craze for collecting plants, especially those from exotic parts of the world, had affected the colonies as well as the mother country. Natal showed more enthusiasm for this phenomenon than did the Cape. E. Percy Phillips has noted of this period, 'The romance of the Cape had passed and more and more attention was being given to the floras of the recently explored countries'.[16]

In the Cape itself the excitement engendered by the creation in 1858 of the post of colonial botanist, first filled by Dr Ludwig Pappe, had burnt itself out by 1866. In that year the Cape parliament abolished the post and the accompanying chair of botany at the South African College was not filled.[17] Yet it was to the Cape that Kew looked for help in the mid-1860s. In 1867 Joseph Hooker wrote to the famous botanist Harry Bolus that the Cape was 'a colony which I think boasts of more [botanists] than any other'.

Two years earlier Sir William Hooker in his last *Report on the Gardens* had noted that Kew's old Australian and South African plants were 'worn out'.[18] To help rectify this matter he wrote to the official colonial botanist at the Cape, Rev. Dr John Croumbie Brown, who had already appealed successfully to missionaries to send plant specimens to Kew. These donors had included Rev. R. Robertson of Zululand. In February 1866 Brown issued a circular in which he stated that Kew was particularly desirous of succulent plants, bulbs and orchids. He quoted the elder Hooker as saying, 'We have not a dozen stapelias left in the garden'. Brown observed that only an estimated tenth of South African species had found their way to Europe in a living state. He rightly commented on the necessity of sending different plants of the same species to Europe and outlined the benefits gained through the 'healthful mental occupation' of plant collecting. The scarcity of South African plants at Kew was in fact largely solved from outside South Africa when, on his retirement in 1873, William Wilson Saunders moved from his Reigate home and donated to Kew the 'unrivalled collection' of South African plants which he had assembled there.[19]

16 Phillips, 'Historical sketch', p.52.

17 Conrad Lighton, *Cape floral Kingdom*, (Cape Town, 1960), p.32.

18 *Kew Report*, (1869).

19 Britten and Boulger, vol 28, (1890), p.279; and *Kew Report*, (1873).

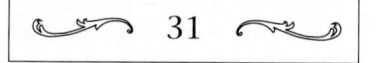

ALBUCA NELSONI N.E.Brown of Kew, who described this member of the family Liliaceae, noted that it was 'the first species of *Albuca* hitherto made known'. It was discovered near the Umlazi River in Natal by the famous commercial plant hunter William Nelson. He sent it to his father's nursery outside Rotherham in England where it first flowered in 1880. Its habitat is rocky outcrops and grassy slopes. (*Curtis's Botanical Magazine*, T6649. [1882])

---

20 K.A.: South African Botanists, 1859-96, miscellaneous reports, vol 12.6.

21 Phillips, 'Historical sketch K, p.40.

22 K.A.: Signature illegible, 1 July 1881, (written from Algoa Bay; South African letter vol 189).

23 For example, in 1869 the Caen Linnean Society published 'Note sur les/lichens de Port Natal'.

24 Keit letters in the possession of the Keit family, Pretoria: W. Keit to Wilhelm Steinmetz, 3 December 1882.

25 *Curtis's Botanical Magazine*, (1882), t6649.

Near the end of his interesting and lengthy 1866 circular the Cape's colonial botanist observed, 'Before leaving England it appeared to me that much more was known there of the products of Natal than of the products of Cape'.[20] This was an astute observation, for the fortunes of Cape botany were already on the decline by the time Brown left the Cape in January 1867, and continued in this way until the 1890s.[21] By 1881 one irate correspondent in the eastern Cape complained bitterly to John Baker of the Kew herbarium, 'Hardly anyone makes any effort to forward the country botanically but merchants import brandy and gin in enormous lots'.[22]

No such remark could be made about Natal botany at that time, which was now being written of even in continental Europe.[23] A growing awareness of Natal's diverse flora crystallised in public concern about the fate of the colony's indigenous forests, in the establishment at Pietermaritzburg in 1874 of the colony's second botanic gardens, and in voluminous correspondence between Natal and Kew.

In December 1882 William Keit, former curator of the Durban botanic gardens, wrote to his cousin in Dresden, 'Explorers of nature come and go and we see and hear little about them'.[24] Many of these individuals were employed on a freelance basis by overseas nurseries and such business concerns as Mr Bull's Establishment for New and Rare Plants in London. One of these commercial plant hunters was the celebrated William Nelson. He collected for his father who ran a nursery near Rotherham in England. *Albuca nelsonii* which Nelson discovered near the Umlazi river was named by Kew in 1882.[25]

Even more illustrious than William Nelson was the intrepid Victorian lady traveller and plant painter Marianne North. Her close links with Kew made her visit to the Cape and Natal in 1882 of special significance, coming as it did at a time, as will be seen, when Natal botany was experiencing a serious crisis. Sir Joseph Hooker was able to get a firsthand account from her of botanical developments in the colony.

Marianne North stayed with, among others, Bishop Colenso, whom she did not like. None the less she painted a fine landscape, complete with goliath heron, of the view from Colenso's home, Bishopstowe. This was one of the 16 oil paintings she did to illustrate the flora of Natal. They have a greater depth and richness than the contemporary paintings of Natal's flora by Mrs Katharine Saunders. The North Natal paintings were soon housed with the rest of her 832 canvasses in the purpose-built North Gallery, opened in June 1882, which she had donated to Kew. In the preface to the gallery's first catalogue Sir Joseph Hooker made a prescient observation concerning the paintings which has often been quoted. What he said was as true of Natal as it was of the many other regions of the globe which Marianne North visited. Her paintings, Joseph Hooker said:

depict scenes many of which are already disappearing, or are doomed to disappear, before the axe, and the forest fires, the plough, and the flock of the ever-advancing settler and colonist. Such scenes can never be renewed by Nature, nor when once effaced can they be pictured to the mind's eye, except by means of such records as these.[26]

The considerable increase in the number of active local botanists in Natal from the mid-1860s was seen by Kew as a desirable development even if it did produce an extraordinarily large number of self-opiniated correspondents to be dealt with. These included James Walker of Richmond who used a schooldays' acquaintance with Joseph Hooker to elicit help from Kew,[27] and the colourful Colonel J.H. Bowker, formerly of the Cape frontier mounted police and ex-chief commissioner on the diamond fields, who was the brother of the Cape botanist Mary Elizabeth Barber. He was an enthusiast of all things relating to nature and was particularly keen on collecting butterflies. Marianne North recounted that he was known everywhere as Colonel Butterfly and that when photographed he always posed with his net, 'an appendage to which the poor Empress Eugenie objected not unnaturally, when it appeared conspicuously in front of a view he presented to her of her son's grave'.[28]

Some 'metallic stone', which he found in Natal, the Natural History Museum in London dismissed as 'bauxite ore, no value'. To Kew he sent the creeper *Grewia caffra* from which he said good walking sticks were made in the colony. He boasted of his links with Sir Bartle Frere and General Colley and in 1880 expressed a wish that Kew inform him how to join the Royal Society.[29]

Another correspondent had the effrontery to lecture Joseph Hooker on the correct manner in which coniferous seeds should be transported and reprimanded Kew for failing to take his advice. This was A.S. White who farmed at Fundisweni in the area between Mtamvuma and Mzimkulu rivers which had been annexed to Natal in 1866 and named Alfred County. With its tracts of yellowwood forest, it proved an exciting place for the botanically minded settler to explore. Interestingly a dominant species on White's farm was the beautiful Natal bottlebrush *Greyia sutherlandii*. In response to Brown's 1866 circular, via the Cape botanist White despatched to Kew some club mosses and several new species including *Mondia whitei (Chlorocoden whitei)* which was named in his honour. The root of this forest plant had a pleasant taste and could be used as a stomachic. It had become scarce because of the amaPondo liking for what they called the 'mundi' or 'Umundi' plant.[30]

Katharine Saunders (1824-1901)
This strong-willed Victorian lady was to make a name for herself not only as a plant collector but also as a botanical artist. From the mid-1850s she lived on an estate at Tongaat just north of Durban. Her son, Sir Charles Saunders, was also interested in plant hunting.
(Courtesy: National Botanical Institute, Pretoria)

26 Allan, p.235
27 K.A.: James Walker, 18 July 1871 and 21 May 1872.
28 Marianne North, *Recollections of a happy life*, (London, 1892), II.274.
29 K.A.: Bowker, 4 May 1880, 17 October 1882 and 17 February 1883.
30 K.A.: White, 15 October 1866 and 23 September 1867; and *Curtis's Botanical Magazine*, (1869), t5898.

M.N.351

M.N.370

The Natal Paintings of Marianne North

Marianne North (1830-90) was an English lady from a reputable family. She had two great enthusiasms in life - travel and painting. Between 1871 and 1884 she combined these hobbies by taking a number of trips to far-off places such as Jamaica, Brazil, North America, Japan, the East Indies, Ceylon, Australia, New Zealand and Chile. In 1882 she came to the Cape and from there she travelled to Pondoland and on to Natal where she stayed from 12 April to 22 May 1883. A brief account of her South African adventure appeared in her memoirs, *Recollections of a Happy Life* that was published in 1892. Her paintings, 848 in total, from her adventures, including those of South Africa of which 16 were painted in Natal and are reproduced here, are housed in the Marianne North Gallery in the grounds of Kew gardens. This building was erected at her own expense and opened on 9 July 1882. On her return from South Africa she had an annex added to the gallery where her later paintings, including those from Natal, are housed.

Detailed captions on page 39

M.N.364.

M.N.354.

◀ M.N.381. 
▶ M.N.355.

◀ M.N. 357.
▶ M.N.368.

M.N.366.
M.N.362.

M.N.386.
M.N.388.

M.N.389.

M.N.383.

M.N.371

M.N.351 **View of the Mountains from the Railway between Durban and Maritzburg, Natal** This view is from the highest part of the railway and the impression of the space and height of the area is reinforced by the purple mountains in the distance reaching to meet the blue sky. The undulating foreground is detailed with cycads and lilies.

M.N.352. **Clivia miniata and Moths, Natal** Marianne North revelled in the colour of her tropical travels and enjoyed reproducing it, as her paintings show. Here the delicacy of the moths is a foil for the golden-oranges and deep reds of the flowers and fruit. With characteristic attention to scientific detail she has shown the fruit in different stages of ripening.

M.N.354. **White Convolvulus and Kaffirboom, painted at Durban, Natal** The convolvulus Marianne North named as *Ipomoea ventricosa*, is a native of the West Indies. The Kaffirboom (*Erythrina caffra*) is now called the coast coral tree. The scarlet, white, green and blue of this picture captures the exhilarating contrasts of a fine, fresh morning in Natal.

M.N.355. **Morning Glory, Natal** The stunning blue *Ipomoea rubrocoerulea*, a Mexican species, which covered 'all the verandahs in Durban... never saw it more lovely', she wrote.

M.N.357. **Blue Lily and Large Butterfly, Natal (357)** Behind are large tufts of *Agapanthus umbellatus*

M.N.362. **White and Yellow Everlasting (with Varieties of Mantis to Match) and other Natal Flowers** This montage of Natal flowers includes the yellow-flowered *Senecio macroglossus*, a climber with ivy-like leaves; the clustered, wholly yellow *Helichrysum appendiculatum* as well as the *Oxalis semiloba* and a species of *Cyrtanthus*. Marianne North noted that the similarity in colouring of flowers and insects helped to camouflage and protect the insects from their 'feathered foes'.

M.N.364. **View of a Table Mountain from Bishop Colenso's House, Natal** Colenso lived at Bishopstowe outside Pietermaritzburg. The eucalyptus and Indian bamboos were planted by the bishop. The goliath heron was a pet. Marianne North found Colenso 'a thorough gentleman' though supposed him weak and vain and very susceptible to flattery'. Like Sir Joseph Hooker she was not impressed by the Colenso 'Zuluism'. 'It would have driven me mad to stay long in such a strained atmosphere', she wrote.

M.N.366. **A Cycad in Fruit in Mr Hill's Garden, Verulam, Natal** Marianne North said this was a *Cycas circinalis*, a cycad of the 'tropics of the Old World'. She wrote that Mr Hill's garden contained many beautiful things.

M.N.368. **Two Flowering Shrubs of Natal and a Trogon** The two flowers she lists as the red *Schotia speciosa* and the white *Gardenia thunbergia*. The bird is the usually reclusive Narina Trogon, its scarlet and green plumage echoing the glossy green leaves and the red *Schotia* flowers in this composition and giving it camouflage in the forest.

M.N.370. **A Tree Euphorbia, Natal** This is the *E.grandidens*. Marianne North mentions passing five round-topped *Euphorbias* when travelling by train from Durban north to Verulam. The yellow-flowered tree gives a splash of brightness, while the little black girl carrying a bundle of firewood on her head puts the vastness of the view into perspective.

M.N.371. **Group of Natal Flowers** Marianne North lists these as *Loranthus natalensis*; blue *Pychnostachys reticulata* and the white-pink *Dombeya burgessiae*. Central is a *Cyrtanthus* with a pale purple species of *Hypoestes* on the right, a wild gourd (*Cephalandra palmata*) and an *Ipomoea* entwining the red clay African pot. The painting is an exciting version of a typical, rather static, Victorian flower composition.

M.N.381. **The Knobwood and Flowers of Natal** This tree is a *Zanthoxylum capense*, the detached knobs of which, according to Marianne North, were used by children as playthings. On the trunk is a tuft of the epiphytic orchid *Angraecum saundersiae*. The other flower is a *Ceropegia sandersoni*.

M.N.383. **A Remnant of the Past near Verulam** These Baines' **Aloes** were in the region of 13 metres high when Marianne North painted them in 1883. In her memoir she recorded: 'Mr H (Hill) sat and watched me at work, much pleased to see his dear aloes at last done justice to. He said not even Mrs S (Saunders) had been to see them, and when he wrote a description of them to Kew, they had coolly asked him to cut one down and send them a "section" for the museum!'

M.N.386. **Aloes at Natal (386)** Marianne North has captured the detail of the leaves and flowers but as with the knobwood (no.381) the portrayal of the trunk seems to have escaped her.

M.N.388. **Various Species of Hibiscus, with Tecoma and Barlaria, Natal** Marianne North lists the flowers in this cottage garden scene as: (central) the yellowish-white *Hibiscus cannabinus;* (top left) *H.surattensis* and (bottom left) *H.calycinus* and one small pale-yellow flower of *H.trionum* amid the blue *Barlaria* on the left. The pale-rose flower on the left is *Pavonia mutisii*. At the top right is the handsome scarlet-flowered *Tecomaria capensis*.

M.N.389. **Cycads, Screw-pines and Bamboos, with Durban in the distance** This painting was done in the botanic gardens in Durban in 1883 when the gardens were some 32 years established. The previous year Medley Wood had become curator. With the possible exception of the *Stangeria eriopus* in the foreground these plants would have been put in by William Keit or Mark McKen. Cycads are still grown on this site. The headland in the distance is the Bluff under which the harbour entrance to Port Natal is situated. Monkeys still inhabit the surviving bush on the hills around Durban. This painting gives a good impression of the distance from the gardens to the town at the edge of the bay.

HAEMANTHUS KATHERINAE (Catherine Wheel) This dramatic member of the Liliaceae family was first sent to Kew as a dried specimen by John Sanderson. In 1887 William Keit, the curator of the Durban botanic gardens, sent a living plant with the request that it be named after Mrs Katharine Saunders. This was duly done, though her Christian name was misspelt in *Curtis's*. It was described by N.E.Brown in the *Gardeners' Chronicle*. (*Curtis's Botanical Magazine*, T6778, [1884])

31 K.A.: K. Saunders, 8 June 1882.

32 K.A.: K. Saunders, 1 April 1881, 26 April 1881 and 16 May 1881.

33 See K.A.: K. Saunders, 6 August 1881; and *Flower paintings of Katharine Saunders*, (Tongaat, 1979).

34 K.A.: Keit, 17 March 1877 and 24 December 1877.

35 K.A.: K. Saunders, 13 March 1882, 3 June 1882 an attached Kew memorandum, 11 July 1882.

36 K.A.: K. Saunders, 11 October 1882.

Even more importunate than any of the above was Natal's premier resident lady botanist, Mrs Katharine Saunders. She was the wife of James Saunders, popularly known as 'the Tongaat slasher', a sugar farmer and member of the Natal parliament. She had a talent for plant drawing and had been much encouraged to collect and dry plant specimens by Mark McKen, 'my first instructor in the botany of Natal',[31] when for a while he had been her husband's estate manager during the break, from 1853 to 1860, between his two terms as curator of the Durban gardens. In 1881, on a visit to Britain and continental Europe, Mrs Saunders met Sir Joseph Hooker and sent him some of her dried specimens on her return to Natal.[32] So pleased was she by the number of novelties detected among these that she passionately devoted the remaining 20 years of her life to plant collecting and painting.[33]

William Harvey, with whom she corresponded, named a new *Habenaria* in her honour. In common with several other Natal botanists Mrs Saunders believed that the naming of a plant in one's honour was the finest accolade a collector could attain. As early as 1877 she had got the curator of the Durban botanic gardens to persuade Kew to do this for her.[34] Kew obliged and named in her honour a beautiful plant in the Amaryllidaceae, *Haemanthus katharinae*. She was not content with this, however, and in February 1882 she asked Sir Joseph Hooker if her son Charles Saunders, a colonial official and collector, might have a novelty named in his honour. More was to come.

A list of the names of some unidentified dried specimens which Katharine Saunders had given to Hooker was sent to her by the Kew herbarium. It included one plant listed as 'orchid genus'. Katharine Saunders asked Kew if it might 'be named in such a manner as to immortalise myself and my son by giving the genus our surname'.

Sir Joseph Hooker, as always when dealing with amateur botanists, was enthusiastic and encouraging in his reply. He passed the matter over to Thiselton-Dyer, who in turn passed it to Professor Oliver in the Kew herbarium with the following note:

Her husband is a member of the legislative council so we must bear with her. Her passion is to have something named after her. Can this be managed?

Oliver, however, was not prepared to be amenable. He curtly told Thiselton-Dyer that such a thing could only be 'earned in the usual way' and that the proposed new 'doubtful' genus was only a 'scrap not identifiable'.[35] When Katharine Saunders heard of this she was not pleased and wrote to Sir Joseph Hooker of her disappointment at not having been commemorated 'as you kindly seemed to agree to do'.[36] But the director of Kew himself could not dictate to the formidable Professor Oliver, even if he had wished to.

In a letter to Kew in the early 1880s, after pointing out that she could generally detect a novelty 'at a glance', Katharine Saunders made mention of a local botanist, John Medley Wood. In matriarchal fashion she stated that a few years earlier Wood 'did not know nearly so many plants we passed on the road as I did!'[37] Wood was later to become the doyen of Natal botany and a strong claim may be made for his being the father of modern Natal botany. Less forward than Katharine Saunders, Wood was none the less not without vanity. A self-trained botanist, he had followed his father to Natal in 1852 and started farming experimentally at the mouth of the Umhloti river. This and the marriage of his younger sister Margaret to Mark McKen had led him to take an interest in botanical matters.[38]

In 1868, at the age of 41, Wood moved inland for health reasons to nearby Inanda where he ran a boarding house and trading store.[39] It was here that he became engrossed in botany, especially in the study of cryptogams.[40] This inevitably led him into a correspondence with Kew which commenced in 1875 and ended only with Wood's death 40 years later. It was Natal's colonial secretary, Napier Browne, who referred Wood to Joseph Hooker.[41] Wood's first letter to the director was uncharacteristically timid.

In a remarkably short time Wood had made a name for himself as an authority on botany. Colonel Bowker, John Sanderson, Robert Jameson, the veteran committee member of the horticultural society, and the visiting Polish botanist Anton Rehmann all came and stayed with Wood. Rev. John Buchanan also befriended him. Buchanan sent many specimens to Kew via McKen and in 1875 he published a 30-page pamphlet entitled *A revised list of the ferns of Natal*. In the mid to late 1870s, before Buchanan left Natal to take up a living in Madeira, he gave Wood much useful guidance.

In his collecting Wood was occasionally helped by Walter Haygarth, the nephew whom he had adopted, a young man who was soon to make his name as a botanical artist.[42] Wood searched the Inanda area and travelled down the coast and inland to the Noodsberg mountains to find specimens. In 1877 he published a 40-page booklet with the cumbersome title, *A popular description of the Natal ferns designed for the use of amateurs*. Two years later this was followed by a 12-page pamphlet, written jointly with Buchanan, entitled *The classification of ferns*. Throughout these years the volume of material which Wood despatched to Kew increased. It consisted principally of herbarium specimens. By June 1879 he had sent 560 such specimens and a year later the figure stood at 934.[43] Ferns, mosses and fungi were his special delight but he was easily sidetracked.[44] In May 1879 he assured Kew that he would 'not neglect any succulent plants that may fall in my way'.[45] Earlier that year he had commenced what proved to be a lengthy correspondence with Kew on what Wood felt was a new species of *Gerrardanthus*.[46]

Kew tactfully trained him, politely advising and recommending and sometimes asking his opinion.[47] They were not slow to retort if they felt Wood was wrong. To his comment that a fern was merely a variety of *Trichomanes pyxidiferum* Kew responded, 'Yes, but a slightly different variety from what we have from Natal already'.[48]

Occasionally they gently reprimanded him: in particular Kew was concerned over Wood's faulty system of numbering specimens. Frequently a listing giving two specimens with the

37 K.A.: K. Saunders, 8 June 1883
38 Gunn and Codd, pp. 379-81.
39 K.A.: K. Saunders, 8 June 1882.
40 See A.M. Bottomly, 'An account of the Natal fungi collected by J. Medley Wood', *S.A.J.S.*, 13, (1916), pp 424-46; and M.C. Cooke and C. Kalchbrenner, 'Natal fungi collected by J.M. Wood, Inanda', *Grevillea*, X, (1881), p.26-7.
41 K.A.: Wood, 15 August 1875.
42 See K.A.: Wood, 13 January 1882.
43 K.A.: Wood, 23 June 1879 and 4 June 1880.
44 The remains of Wood's Inanda collection of ferns, numbering 104 sheets, may be seen at the Natal Herbarium, Durban.
45 K.A.: Wood, 9 May 1879.
46 This correspondence on the *Gerrardanthus* begins on 14 January 1879.
47 Natal Herbarium, Kew list volume, February 1882.
48 K.A.: Wood, 29 August 1879.

This letter dated 1873 from the artist, explorer and botanist Thomas Baines to Sir Joseph Hooker was written from the Temperance Boarding House in Durban. It discusses several matters including specimens of an *Aloe* he was sending to Kew. This proved to be the giant tree *Aloe* which was appropriately named *Aloe bainesii*.
(Courtesy: Royal Botanic Gardens, Kew)

49 Natal Herbarium, Kew 1st volume, October 1879, January 1880, April 1880 and August 1881. See also Ross's *Flora of Natal*, pp.4 and 202.

50 K.A.: Wood, 23 July 1879.

same number proved to be two distinct species when examined in the Kew herbarium.[49] Wood was not always amenable to such criticism. In July 1879 he stiffly defended himself, 'I gathered first the flowers and afterwards the ripe seed vessels and sent both together - but both specimens were got from the same plant'.[50] It was impossible for Wood to defend himself against some of the remarks flowing from the pen of Thiselton-Dyer, Oliver and Brown. In 1883, for example, the Kew herbarium sent Wood the following observation:

I think it is a pity to number different gatherings the same, unless they are positively from the same plant: That plan has caused heaps of confusion in Herbaria. The only really reliable numbers are Burchell's. He never gave different gatherings the same number although he often knew them to be the same plant, so there is never any fear that the same number of Burchell's in different Herbaria may represent different plants.[51]

Very occasionally Wood did have the satisfaction of seeing Kew make a mistake. Concerning specimen 292 Kew briefly noted: 'If I sent you the name *V. marginata* for this before, it was by a slip of the pen'. Considerably more satisfying was the recognition Wood was receiving from Kew as his name was attached with increasing regularity from the early 1880s to plants described in *Icones Plantarum*.[52]

A sketch of some tree aloes (*Aloe bainesii*) growing on the lower slopes of the Drakensberg mountains. (From, *The Gardeners' Chronicle*, 2 May 1874)

THIS NEW GENERATION of Natal botanists had benefited greatly from the work of the early pioneer collectors. They were generally more knowledgeable about the flora of the colony and very much more brash about proclaiming their knowledge. Apart from those letters from the older collectors, Kew now only occasionally received correspondence which was apologetic in its respectfulness and obviously written by someone who held Kew in the greatest awe. One such correspondent was the artist Thomas Baines, who was a friend of McKen and who wrote to Kew from the temperance boarding house in Durban concerning the desirability of introducing English wheat into Mashonaland. He also discussed his paintings with Kew.

Though 18 species carry Baines' name, they are nearly all from the interior of the continent rather than Natal. Baines did, however, send Kew samples of 'cotton' growing on trees in the Tugela valley and in 1873 he sent Kew sketches of an aloe he had found near Inanda as well as one of the plant's large branches. This species eventually came to be known as the great tree aloe, *Aloe bainesii*.[53] Kew had heard of this aloe through McKen as early as August 1866, when he described one 30 feet (10 metres) high growing on the banks of the Umhloti river. Sir James Hulett had also sent a description of several at Verulum to Kew, who had 'coolly asked him to cut one down and send them a "section" for the Museum!'[54]

51 Natal Herbarium: Kew list volume February and July 1883.

52 An early example of Wood's growing recognition may be seen in *Niebuhria woodii*, described in *Icones Plantarum*, 3rd series, vol IV, (June 1882), p.67.

53 See K.A.: Baines, 3 January 1869, 22 July 1872 and 15 July 1873; Keit, 16 July 1873; and Sanderson, 1 July 1872; and Allan, p.214.

54 *A vision of Eden, The life and work of Marianne North*, (London, 1980), p.216.

43

Chapter 4

Botanical institutions develop (1866-83)

THE INCREASED POPULAR INTEREST in botany in Natal benefited the development of its botanical institutions. A botanic society was founded in Pietermaritzburg at the end of 1872 and in March 1874 the city corporation granted it 101 acres (41,1 hectares) of land for the purposes of establishing a new botanic gardens in the colony. Like the Durban gardens it was given an annual grant of £350.

Even though in 1865 McKen of the Durban botanic gardens had recommended the establishment of a branch gardens in the colony's capital city, a recommendation which had been unacceptable to Pietermaritzburg,[1] when the new botanic society determined to establish its botanic gardens one of its members noted, 'We are sure to have to fight with the coast about it...for they think sugar and coffee should take everything'.[2] Because of the considerable difference in climate between Durban and the capital, some 50 miles (80 kilometres) inland and 2 200 feet (670 metres) higher than the coast, the Durban gardens had not proved terribly successful in experimenting with crops suitable for growing in the agriculturally rich Natal midlands. As it was, the new botanic gardens, some two miles (3 kilometres) distant from the city centre in the steeply sloping Zwartkop valley, proved itself to be extraordinarily successful in raising exotic seedling timber and fruit trees for distribution either free or at a nominal sum. In this respect it served the same purpose as the Grahamstown botanic gardens in the eastern Cape. By 1880 the Pietermaritzburg botanic gardens could offer the growing number of tree plantation owners five kinds of *Eucalyptus*, five of pine, three of cypress and two of *Acacia*, including the Australian black wattle *(Acacia mearnsii)*, the production of which was to bolster the ailing Natal economy.[3]

Because of this specialisation in seedling production the Pietermaritzburg botanic gardens had little contact with Kew in the Victorian period. This did not mean that Kew was unaware of its existence or of what was happening there. Indeed Sir Joseph Hooker considered the Pietermaritzburg botanic gardens to be the finest seedling-producing institution in the British empire. Referring to it in 1884 Thistelton-Dyer, quoting Hooker, commented a promoting afforestation, 'Natal may be held up as a model to other colonies'.[4]

THE ESTABLISHMENT OF a botanic gardens in Pietermaritzburg did not result, as some had predicted, in the decline of the Durban gardens. Following a stiff reprimand from the secretary of the Natal Agricultural and Hor-

---

1 K.C.A.L.: Durban Botanic Society, KCM 43065, Memorandum book 1, 1865-71.

2 K.A.: Topham, 28 November 1872 and 15 September 1874; and Sutherland, 24 February 1873 and 31 August 1874.

3 K.C.A.L.: Pietermaritzburg Botanic Society curator's report, 1880.

4 K.A.: *Correspondence and reports relative to the state of botanical enterprise in Natal, 1882,* (Government report, Pietermaritzburg, 1884), p.2.

ticultural Society to McKen in March 1865,[5] much was done to put the gardens in order: a small greenhouse, two summerhouses, two ponds and a 3 000 gallon (13 600 litre) water tank were constructed in the gardens.[6] The gardens were still not properly laid out though they did contain some very interesting exotic flora,[7] including the much admired crimson *Antigonon leptopus* which had only recently been introduced into Europe from Mexico.[8] Indeed McKen became very selective about what he would allow to be planted in the gardens and on one occasion he ventured to grumble to Kew about their having sent him azaleas: 'I beg to say that other plants would have been preferred'.[9] He did very little to label any of the plants but took delight in personally informing special visitors of all the treasures the gardens contained. Such visitors included Joseph Hooker's nephew, Malcolm McGillvary, and the famous director of the Pamplemousses botanic gardens on Mauritius, Dr Charles Meller.[10]

McKen himself travelled to Mauritius in 1867 and spent seven weeks collecting there. He also continued to collect in Natal and had by 1868 a large number of 'friends up country' who collected for him. They were as far afield as 'No-man's-land' on the Drakensberg escarpment.[11] Most, but not all, of these specimens were despatched to Kew: in 1870 McKen mentioned to Joseph Hooker that he was going to make a tour upcountry to get some *Encephalartos* 'for Queensland and Sydney'.[12] In 1866 McKen recorded that he exchanged plants with the botanic gardens in Adelaide, Cape Town, Ceylon, Kew, Liverpool, Mauritius and Melbourne, as well as with 22 foreign individuals and private firms.[13]

Increasingly McKen was drawn into the Victorian craze for fern collecting, or pteridomania.[14] In this he found a partner in Rev. John Buchanan, Medley Wood's friend. Jointly Buchanan and McKen collected ferns and sent them to Kew. They even sent some fern fossils for the Kew museum section.[15] In return Kew sent McKen a copy of *Synopsis of all known ferns*.[16] In this McKen marked 115 species which he had 'In my Herbarium'.[17] In 1866 McKen sent the Kew herbarium a complete set of his dried ferns. He urged all he knew to collect ferns and in July 1868 wrote to Hooker, 'Mr Baker will be glad to know that nearly all here are fern hunting'. Many of these specimens were added to Kew's collection of 800 species housed in the new fernery.

McKen had hoped to bring out a flora of flowering plants in the Durban area but in fact it was to be in the field of ferns that he published. In 1869 he published his short but good *Ferns of Natal* which was given a 'kind notice' by Kew. A year later McKen collected together material he and the late William Gerrard had gathered and brought it out under their joint names as *Synopsis Filicum Capensium*.[18]

IN ADDITION TO THE SPECIMENS which McKen and Buchanan had sent to Kew, McKen acted as a forwarding agent for most of the collectors in the colony and sometimes even beyond: the Portuguese consul in the Transvaal used McKen to forward specimens to Kew[19] as did McKen's friend Edward Button, who was to be commemorated in the genus *Buttonia* that McKen had named after him. The jovial Button wandered around south-east Africa in search of gold and plants, both of which he found. He appears to have been excited as about discovering in Natal a 30-foot-high (9 metres) *Encephalartos* 'with three heads' as he was about his discovery of the Eersteling gold reef near Marabastad.[20]

The practice of sending McKen

5 K.C.A.L.: Durban Botanic Society, KCM 43065, Memorandum book 1, 1865-71, Vause to McKen, 7 March 1865.

6 K.A.: McKen, 22 June 1868 and 18 May 1871.

7 See curator's reports, 1867-1871, contained in *Natal government gazette*.

8 K.A.: McKen, 10 August 1866 and 8 March 1870.

9 K.A.: McKen, 10 March 1869.

10 K.A.: McKen, 13 September 1871 and 12 October 1871.

11 K.A.: McKen, 18 August 1868.

12 K.A. McKen, 16 August 1870.

13 The full list is in the abovecited Durban Botanic Society memorandum book 1, housed in K.C.A.L.

14 See D.E. Allen, *The Victorian fern craze*, (London, 1969).

15 K.A.: McKen, 20 April 1867, 11 July 1867 and 14 December 1868; and K.A.: Plant list volumes X.282-3 and XVI.223-33.

16 K.A.: McKen, 14 December 1868.

17 McKen's copy of *Synopsis of all known ferns* is still housed at the Natal Herbarium.

18 See K.A.: McKen, 18 October 1896 and 22 February 1870.

19 K.A.: Mcken, 21 November 1871.

20 K.A.: McKen, 10 February 1867 and 11 July 1867; Natal Colonist, 3 February 1871; and Icones Plantarum, 3rd series, vol 1, (London, 1867-71),pl. 1080.

specimens to forward to Kew worked well, for wardian cases could be stored at the gardens until full. But occasionally this procedure caused problems as in Kew's records these cases were usually recorded under McKen's name. This caused considerable irritation to competitive collectors who were possessive of their finds. John Sanderson on several occasions had to inform Kew, 'I filled it [the wardian case], McKen contributed only a few things'.[21] A.S. White accused McKen of taking the credit and scientific name of 'the peanut tree' which White had discovered and sent to McKen for forwarding to Kew. Later, when White discovered what he thought was a new *Zamia*, 'like the one the King of Prussia got from Japan', he told Kew that he would not be cheated by McKen this time but would give him 'a wide berth'.[22]

McKen was sometimes in trouble from Kew as well. In 1870 he was severely reprimanded for having sent a nest of live insects with some plant specimens.[23] But on the whole Kew staff were tolerant of McKen's weaknesses and recognised his respect for Kew.[24] This was indisputable: should two or three months pass without Kew writing to him, McKen would make it known that he would like 'the pleasure of a letter from you'. He was also keen to have photographs of Joseph Hooker and his late father.[25]

ON 20 MARCH 1872 John Sanderson told Joseph Hooker 'McKen is very poorly and can't write'. Exactly a month later McKen died, aged 48. Like many Natal plant collectors he had suffered from malaria. In addition, according to Sanderson, he drank too much and died of an enlargement of the liver, having been administered opium.[26] P.C. Sutherland wrote to Hooker, 'It is a colossal loss and one which will not be soon made up'.[27]

Two years previously McKen had told Hooker, 'I deem it prudent for the sake of my family to "look before I leap"'.[28] Sadly McKen had had little opportunity to be prudent financially. In the 1860s he had lost the right to sell the gardens' surplus stock for his own benefit and in the last years of his life the government grant had been reduced to £250 per annum; it was not restored to £350 until 1873.[29] A relief fund was launched for McKen's destitute wife and six children. A concert to raise funds was held in Trafalgar Hall in Durban on 25 April. Robert Topham told Kew, 'Even at the diamond fields several gave freely'. Joseph Hooker sent a donation. The final sum invested for Mrs McKen, 'a very excellent woman', was £600.[30] In a simple note in the *Journal of Botany* it was recorded:

The Royal Botanic Gardens at Kew are greatly indebted to Mr McKen for living plants and seeds of South African species and our herbaria have been much enriched by the results of his explorations in Natal and the adjacent districts.[31]

Interestingly, two years later the headmaster of Hilton College in Natal wrote to Hooker and mentioned that one of McKen's sons was at the school and that they hoped to 'get him under your eye at Kew'.[32]

During the last years of his life McKen had been the *bête noire* of the president of the horticultural society, John Sanderson. Sanderson lost no opportunity of denouncing the curator to Kew. He accused McKen of stealing society books, of drinking too much, of misleading Kew about who was sending them plants and of generally failing to run the gardens properly. Even after McKen's death Sanderson could not resist condemning his fellow Scot. Only once did McKen complain to Kew of Sanderson, blaming him for packing a wardian case

---

21 K.A.: Sanderson, 11 October 1869.

22 K.A.: White, 8 January 1868, 25 September 1868 and 3 February 1869.

23 K.A. McKen, 12 March 1870.

24 See K.A.: McKen, 11 July 1867.

25 K.A.: McKen, 11 October 1867.

26 K.A.: Sanderson, 22 April 1872.

27 K.A.: Sutherland, 17 May 1872.

28 K.A.: McKen, 15 October 1870.

29 K.A.: McKen, 18 August 1860 and 22 June 1868; and Sanderson, 8 March and 22 June 1868.

30 See *Natal Mercury*, 25 April 1872 and 8 May 1872; and K.A.: Topham, 14 August 1872; and McKen, 15 July 1872 and 19 August 1872.

31 *Journal of Botany*, (1872), p.223. See also *Gardeners' Chronicle*, (1872), p.806.

32 K.A.: Rev. W.A. Newnham, 9 April 1874.

poorly. Indeed McKen seems to have been strangely naive about his enemies. After A.S. White had denounced him to Kew, Kew wrote to McKen asking who this Mr White was. McKen wrote at length of White's high qualities as a botanist and 'excellent observer'.[33]

Sanderson was particularly annoyed when McKen took an interest in orchids. 'He is mad after them', McKen told Kew of Sanderson's special fascination. When Kew sent McKen a collection of West Indian orchids in 1867 Sanderson told Hooker, 'I fear they will be wasted at present'. Four months later he wrote to Kew:

McKen says Gerrard found Eulophia revoluta at the Tugela some years ago and he did, himself, between the Garden and Durban, but without knowing it to be an orchis. I have my doubts about this but it may be so.[34]

While this petty feud was undesirable in Kew's eyes, they managed to remain remarkably neutral and thus prevented its escalation. With McKen supplying them with valuable ferns and Sanderson sending numerous orchids, the two principal figures associated with Durban botanic gardens were sending Kew the most sought after plant specimens in mid-Victorian Britain.

Sanderson took over the mantle of the colony's resident orchid expert from the Irishman George Fox Fannin, of the farm Dargle, who had died in 1865. Not only did Sanderson collect epiphytic and terrestrial orchids but he also sketched them. In the late 1860s he sent his sketches to Joseph Hooker in the hope that Kew would arrange for their publication. When Hooker returned them Sanderson was bitterly disappointed and as late as 1878 he was still reminding Kew of their existence.[35] After his death in March 1881 Sanderson's botanical illustrations were offered to the Durban town council which turned them down. His widow showed them to Marianne North in 1882 and she was impressed by them. Eventually the collection was sold to Kew for £8.[36]

Despite the setback to his artistic hopes suffered by Kew's initial rejection of his sketches, Sanderson continued to study orchids and his letters to Kew are full of authoritative references to them.[37] On 5 March 1868 he sent Joseph Hooker a list of the orchids he had collected and sent to Kew. As well as 12 unnamed specimens he listed the following:

Melanea : *Liparis* and *Bulbophyllum*;
Vandea : *Eulophia, Lissochilus, Cyrtorchis, Angraecum, Mystacidium, Ansellia,* and *Polystachya*;
Orchidaceae: *Satyrium, Habenaria, Stenoglottis, Tryphia, Disa, Monodenia, Schizochilus, Brachycorythis, Huttonaea, Brownleea, Corycium* and *Disperis.*

ANEMONE FANNINII In the Natal midlands there was an enthusiastic amateur botanist called George Fox Fannin (1832-65). His father had brought the family out from Dublin and established a farm which he called the Dargle, after the Irish river. A nearby forest also carried the same name. George Fannin discovered this giant *Anemone* in 1863. It was described by Professor Harvey of Trinity College, Dublin, but he incorrectly named it after "Mrs G.Fannin". In fact it had been mounted by Marianne Fannin, George's maiden sister. Only in 1883 did Kew receive a live specimen from Medley Wood. (*Curtis's Botanical Magazine*, T6958, [1887])

33 See K.A.: Sanderson, 20 September 1866, 7 December 1867, 18 August 1868, 11 October 1869, 15 February 1870, 22 April 1872, 1 July 1872 and 11 December 1872; and McKen, 18 August 1868.

34 K.A.: Sanderson, 7 December 1867 and 7 March 1868.

35 K.A.: Sanderson, 5 March 1868, 18 January 1869, 19 November 1869 and 5 March 1878.

36 K.C.A.L.: file 26944, Jowitt to Campbell, 8 February 1949.

37 For example, see K.A.: Sanderson, 20 September 1866, 21 January 1867, 6 March 1868, 8 June 1868, 20 July 1868, 8 January 1869 and 22 August 1869.

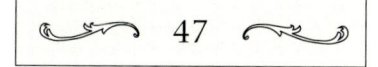

DISA COOPERI The distribution of this member of the family Orchidaceae extends beyond Natal's borders. It was named in honour of Thomas Cooper, the father-in-law of Kew's N.E.Brown. Cooper collected in Natal from late 1861 until September 1862. (*Curtis's Botanical Magazine*, T7256, [1892])

48

Sanderson sent at least 80 different species of orchid to Kew from Natal. These were gratefully added to the ever-expanding Kew orchid collection: from numbering 850 species in 1850, by 1908 it had been built up to 1 800.

The Sanderson family was influential and well known. John Sanderson could boast of a close friendship with Bishop Colenso and Mrs Sanderson was 'an intimate friend of George Elliot'. While being a newspaper editor hampered John Sanderson in his collection, it led him into contact with many diverse types of people and gave him a vehicle, through the *Natal Colonist*, to proselytise in the cause of botany. Sanderson was also a leading figure, with P.C. Sutherland, in the popular Natal Natural History Association, which was established with branches in Durban and Pietermaritzburg in 1868. Joseph Hooker was made an honorary member of the association and was so pleased at the gesture that he offered to give the association specimens as the nucleus of a Natal economic botany museum.[38]

Kew's appointee

IN AUGUST 1908 SIR William Thiselton-Dyer, by then retired as director of Kew, wrote that Sir Joseph had initiated two of the main features of Kew:

These are in vigorous operation at the present day: one was the making Kew the depot for the interchange of plants with the Colonies, "which must prove of great advantage to the commerce of these kingdoms"; the other was the training of young gardeners for botanical and cultural posts abroad...A chain of Kew-trained men dot the course of the future Cape to Cairo railway.[39]

In 1870 Kew began to issue diplomas in horticulture to its successful apprentices. Many requests were made to Joseph Hooker to supply suitable curators for botanic gardens and stations, both in Britain and the empire, and even occasionally to provide head gardeners for important public parks such as Phoenix Park in Dublin.[40] For easy reference Hooker kept a 'List of Applicants', in which the names and details of suitably qualified gardeners were recorded. Not all of these were Kew-trained: if Hooker was impressed by a young man whom he had met, he was prepared to enter his name on the list. It was one such gardener who came to mind when John Sanderson wrote to Joseph Hooker in 1872 asking for a new curator of the Durban botanic gardens following McKen's death.[41]

Sanderson's request was for a young man who was single, out of his apprenticeship, steady, sober and not a man who would bully the African workforce. The man whom Hooker felt

Julius Wilhelm (William) Keit (1841-1916) William Keit worked as a gardener in continental Europe, England and Glasnevin botanic gardens in Dublin before coming to Natal in 1872 on the recommendation of Sir Joseph Hooker. Drought, a labour shortage, lack of funds and a hostility to him as a German made his period as curator (1873-81) a difficult one. Kew stood by him during these difficult years and he in turn sent them what plants of interest he could find in the vicinity of Durban. in later life Keit did well as curator of the city's parks department and as owner of a private nursery.
(Courtesy: Mrs K.Plekker)

38 K.A.: Sanderson, 5 March 1868, 20 July 1868 and 18 January 1869.

39 Bean, pp.xvi and xviii.

40 Allan, p.153; Huxley, II.137; and King, *The world of Kew*, p.21.

41 K.A.: Sanderson, 22 April 1872, 1 July 1872 and 19 August 1872.

HAEMANTHUS KATHERINAE (Catherine wheel). Named after Katharine Saunders but misspelt. (*Curtis's Botanical Magazine*, T6778, [1884])

most deserving of this post was a 31-year-old German called Wilhelm Keit, known as William Keit in the English-speaking world.

Keit had been something of an itinerant gardener, having worked in gardens and nurseries in Dresden, Basle, Paris, Brussels, Nottingham and in the exhibition palace winter garden in Dublin. In May 1868 he had joined the staff of the Royal Botanic Gardens of Ireland at Glasnevin outside Dublin. Here he had risen to be foreman in charge of the propagating houses. The director of Glasnevin, Dr David Moore, referred to Keit as 'an excellent practical gardener'.[42] Hooker had at first thought of appointing Keit to 'another and so much better paid appointment' but he felt that the Natal job would be more suitable for him. Sanderson's request that as well as a curator Hooker should appoint a gardener for the Durban gardens at an annual salary of £50 angered Hooker, who said that to offer such a low sum would expose him to ridicule. The only solution Sanderson could see was for the Durban and Pietermaritzburg botanic gardens to merge, with a single curator and an annual government grant of £700. His prediction that he doubted if 'the Maritzburg people would agree' proved accurate and only the position of curator in Durban was filled.[43]

On 29 August 1872 Keit wrote to Hooker from Glasnevin accepting the Natal curatorship. He admitted:

My botanical knowledge is more general than particular. I can arrange plants botanically after knowing their names, but I could not undertake to know or describe new plants scientifically.[44]

The *Kew Report* for 1872 noted with enthusiasm that 'a skilful superintendent' had gone to Durban gardens. Sanderson was equally pleased with the new arrival who landed on 14 December, though he observed Keit's lack of knowledge of African languages was a slight problem.

The quiet and reserved Keit was rather shocked by his first sight of the gardens. To Dr Moore he wrote:

They are very much neglected, there is no systematic arrangement, the plants having been planted where there was space and, what I regret most, there are no names on them.[45]

None the less Keit made a good start with putting the gardens in order and laying out the pinetum containing 27 different conifers only shortly after Kew had first established its own pinetum. In 1875 the gardens was honoured with a visit from General Wolseley. In his diary Wolseley made mention of Keit's enthusiastically 'pouring into my ear Latin and Greek names of trees and plants until my temper and my legs very nearly broke down from fatigue'.[46] This was in keeping with his jaundiced view of the colony and its people: everybody in Natal he firmly believed had bad breath.

---

42 National Botanic Gardens of Ireland, Glasnevin: Moore to Hooker, 27 August, 1872.

43 K.A.: Sanderson, 11 November 1872 and 11 December 1872.

44 K.A.: Keit, 29 August 1872.

45 Glasnevin: Keit to Moore, 10 January 1873.

46 Adrian Preston, *The South African diary of Sir Garnet Wolseley*, (Cape Town, 1971), p.166.

50

had formally declared its aims to be:

The introduction, cultivation, propagation and distribution of plants of commercial and economic value, the collection and preservation of remarkable plants as objects of curiosity and interest to be still kept in mind but made incidental and secondary to the primary aim.

From then on the gardens increasingly acted as an agricultural experimental station; arrowroot and sugar cane had had some success but not as great as had been expected. The gardens also served as distribution centre for imported 'economics' and, at least until 1881, as unofficial place of quarantine for newly arrived plants.[48]

There was a new feeling of hope in the colony; the old uncertainty about whether the British would 'withdraw', as they had done in the Orange River Sovereignty, had passed and as with Kenya 50 years later there was the feeling that Natal had the potential to make its mark on the empire. As one correspondent pointed out to Kew in 1870, 'Natal with the uplands of the interior must become a grand country for Europeans'.[49] P.C. Sutherland did much in conjunction with Dr R.J. Mann, a former Natal superintendent of schools now residing in London, to keep the name of Natal to the fore at international exhibitions.[50] Even when the colony was virtually bankrupt in the 1860s Sutherland organised the loan from the Kew orangery, which then served as Kew's third museum, of the Natal timbers previously shown at the great London exhibition of 1862. In the early 1880s Walter Peace, the Natal government emigration agent, also tried his best to promote Natal's botanical image and to encourage Kew to assist the colony.[51] Such encouragement was hardly necessary as only rarely was Kew unable or unwilling to attend to a colonist's request.

Though Keit found it difficult to leave his work to collect new specimens, he did take an interest in sea algae and what plants he could find in and around Durban. He sent the first live specimen of *Haemanthus katharinae* back to Kew as well as a new variety of *Littonia modesta* which was to carry his name. In return Kew sent him cases of orchids to supplement the meagre show of indigenous ferns, colensoas and fuchsias that were growing in the gardens' greenhouse.[47]

Kew and the economic development of Natal

WILLIAM KEIT CONTINUED the experimentation with crops which it was hoped might prosper in the colony and stimulate Natal's sluggish economy. Such experiments had been carried out since the gardens' foundation and in 1867 the horticultural society

LITTONIA MODESTA (Butterlily or Yellowlily). This herbaceous rambler with a fondness for forest undergrowth was discovered by John Sanderson. It was described by Sir William Hooker who named it after the late professor of botany to the Royal Dublin Society, Dr Samuel Litton, who like the plant had an 'unassuming and retiring disposition'. It is a member of the Liliaceae family. A variety was named after William Keit. (*Curtis's Botanical Magazine,* T4723, [1853])

47 See *Curtis's Botanical Magazine,* t1884), t6778; E. Charles Nelson and Wendy Walsh, *An Irish flower garden,* (Kilkenny, 1984), pp.83-4 and 183; K.A.: Keit, 23 September 1873.

48 See Donal P. McCracken, 'Natal's botanic gardens: their role in the colonial economy and society', paper presented to the conference of the South African Historical Society, University of Cape Town, January 1985.

49 K.A.: Topham, 19 February 1870.

50 See K.A.: Office of the commissioners for the Paris exhibition, Pietermaritzburg, 20 July 1967; and Sutherland, 8 October 1866.

51 K.A.: Walter Peace, 24 September 1883.

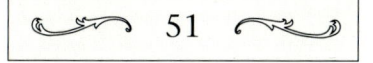

BY THE 1860S LOW international prices had knocked the bottom out of the Natal arrowroot market. Sugar cane continued to be grown but not without problems. The indigenous Natal Green and the many imported varieties, including large consignments imported from Mauritius in 1867 and from Kew in the mid-1870s, had proved unsatisfactory as plantation crops. In 1867 Sutherland was seriously considering the possible alternative of growing sugar beet.[52] Gradually, however, China cane began to assert itself as the most suitable variety even though it too was plagued by a *Cypress* which grew among the young canes.[53] Later, in the early 1880s, this cane was attacked by the fungus *Ustilago sacchari* (cane smut). In 1882 Medley Wood reported to Kew that they should send elephant cane to Natal as the China variety was 'exterminated here'. This Kew did, but it was the Uba variety that was to save the Natal sugar industry.

Equally beset with problems was the successful cultivation of coffee in Natal. Originally introduced from Bourbon in 1854 with more plants being sent later from Ceylon, by the 1870s there was widespread interest among the coastal farmers in the crop despite the heavy losses in coffee plantations through frost in 1869.

In 1874 the disease which had decimated the Indian and Ceylon coffee plantations attacked Natal. To counter this in 1874 and again in 1875 Keit imported and distributed new varieties of coffee plants sent by Kew. Of the coffee plantations Keit wrote to Kew, 'Many an estate has, between Borer, Bark disease, unsuitable soil, climate, and no doubt careless cultivation, probably want of sufficient labour, gone to destruction'.[54]

Kew responded by despatching large numbers of Liberian coffee plants to Natal. Mixed reports were received of their performance but they too were finally to succumb to disease, caterpillars and, in the case of that grown in the Durban botanic gardens, monkeys. In May 1879 Keit reported to Kew that:

If this country is visited periodically by successions of dry seasons...the cultivation of such colonial industries of plants such as Liberian coffee, nutmeg, clove, etc, etc, can not be attempted.[55]

Rumoured yields of a ton of beans an acre made relinquishing coffee cultivation very difficult for plantation owners but finally most were forced to change to sugar cane or to tea production.

Halleria lucida had been used by early colonists as a tea and despite the efforts of Mrs Plant, McKen and a farmer named Jackson it was not until the 1870s that tea planting became popular. While Kew supplied Natal with several consignments of good quality Assam tea plants, both Kew and Keit were nervous about its introduction after the problems encountered with coffee. Hulett and the majority of tea planters brought in plants directly from Indian tea planters and botanic gardens. While tea did not rival cane production, despite its poor quality it was surprisingly successful as a 'mixer' for use in blending in the colony.[56]

Crops which Kew actively encouraged Natal farmers to grow as well as sugar cane were olives, plants of which were particularly in demand from Kew in the mid to late 1860s;[57] cotton, which suffered from red spiders and had to face international competition; and tobacco.

Since McKen had asked Kew for some Shiraz tobacco seeds in 1866 the demand for them had grown. In 1871 the *Kew Report* was able to announce that tobacco grown from seeds which Kew had sent to Natal had been made

---

52 K.A.: Sutherland, 8 May 1867.

53 K.A.: Addison, 9 June 1867.

54 See K.A.: Keit, 6 December 1874 and 21 November 1875.

55 K.A.: Keit, 28 May 1879. See also Keit, 21 January 1874; *Kew Report* for 1874, 1876, 1880 and 1882; and the *Natal Mercury*, 10 December 1880.

56 K.A.: John Bisset, 10 September 1879; Keit, 17 March 1877; McKen, 1 January 1866; and Topham, 18 July 1869. See also Durban curator's reports for 1875 and 1877.

57 K.A.: McKen, 12 June 1869; Sutherland, 23 April 1869 and 23 July 1870; Topham, 18 July 1869 and 19 February 1870; and White, 15 October 1866.

into cigars that were being supplied to the inhabitants of the diamond and gold fields in the interior.[58] By 1874 Sutherland was boasting to Kew that Natal tobacco was 'as good as any we can import'. Natal, here again had to suffer from the problem of a small market and like cotton, tobacco failed to be the saviour of Natal agriculture.[59]

In 1868 Joseph Hooker had written to Harry Bolus at the Cape, 'I cannot fancy what you will do with the chinchonas for which I fear you are too cold and dry. There is not a ghost of a chance of *chinchona* succeeding in South Africa'.[60] It was a prophetic statement but the presence of malaria in Natal and the high profits made from *chinchona*-growing in Ceylon made it an obvious crop with which to experiment. However, seeds of *chinchona succirubra* offered by the Durban botanic gardens in 1874 did not draw the expected response: there were only 32 applicants. By the early 1880s it was clear that *chinchona* was not going to be grown commercially in Natal.[61]

Yet another plantation crop unsuccessfully promoted by the Durban botanic gardens was rubber. In 1875 *Hevea brasiliensis* was sent by Kew to Peradeniya botanic gardens in Ceylon. From here it was distributed to other regions, the most famous for future rubber production being Malaya. In 1878 Peradeniya sent rubber plants to Keit, but a year later he reported to Hooker, 'I fear that this plant will not be suitable for cultivation in this climate'. Though the 1880 *Kew Report* noted that in Natal rubber plants were growing luxuriantly', it was soon obvious that Keit had been correct.

If coffee, chinchona and rubber proved difficult to grow in Natal's coastal sub-tropical region, several other species, sent out by Kew, flourished. These included yams, and varieties of banana and pineapple.[62] Some plants, including bamboo and such fibrous species as hemp, flourished but found little favour.[63]

William Addison, the district surveyor of Victoria County, farmed in the lower Umvoti valley and on his estate had successfully experimented with banana, American aloe (*Agave* sp) and papyrus for making paper, specimens of which he sent to the editor of the *Times* in London. But Addison particularly championed China grass, *Boehmeria nivea*, of which he was able to produce eight-foot-high (2,4 metres) specimens. In 1867 he wrote to Kew of his belief that Natal was fit only to grow fibrous grasses. A problem was the expense of labour in the colony and the need for powerful steam machinery to reduce the plant to a fibrous state. Troubled by the thought of the indigenous 'red nettle' going to waste he told Joseph Hooker, 'I annually destroy tons of [its] fibre in my cane fields'. Having asked how this nettle might be treated he dryly added, 'I presume we must wait till some cute yankie provides the suitable cleaning apparatus'.[64] Despite Addison's proselytising efforts on behalf of China grass, when seeds of it were offered to farmers by the Durban botanic gardens in 1870 there was a poor response.

Forestry

MUCH MORE SUCCESSFUL than such fibrous plants was the growing of exotic trees in plantations in the Natal mist belt of the midlands. Natal was not naturally a forest region but yellowwood forest had existed on the Drakensberg escarpment and along the slopes of the midland hills. Some of this growth, like the great forest of Karkloof, had been badly cut into, the process of deforestation being greatly accelerated by the use of steam-driven sawmills. The agricultural practices of both Africans and

58 *Kew Report*, (1871)

59 Huxley, II.5; K.A.: Keit, 3 March 1875 and 24 December 1877; McKen, 1 January 1866 and 22 February 1870; Sanderson, 8 March 1866; Sutherland, 23 May 1868, 23 July 1870 and 24 November 1874; Topham, 9 August 1870; and James Walker, 21 May 1872. See also the **Natal Colonist**, 24 November 1871.

60 Huxley, II.4

61 See K.A.: Keit, 27 July 1876; McKen, 1 January 1866, 22 November 1866 and 23 October 1867; K. Saunders, 24 August 1881 and 29 August 1881; and Wood, 3 September 1883. See also W.B. Turrill, *The Royal Botanic Gardens, Kew: Past and present*, (London, 1959), p.35; and Natal Government Notice number 131, (1874).

62 See K.A.: Keit, 23 September 1873 and 21 January 1874; and McKen, 1 January 1866, 10 August 1866 and 26 June 1867.

63 See K.A.: Sanderson, 19 November 1869; and Sutherland, 24 February 1873, 26 March 1874, 9 November 1866, 10 December 1866, 23 April 1869 and 24 January 1874; and *Kew Report* (1878). Kew also took an interest in South African bamboo: see *Kew Report*, (1878); and K.A.: Wood, 4 June 1880 and 13 January 1880.

64 K.A.: Addison, 9 June 1867.

colonists further reduced the forest size, as did the heavy demand for timber for vehicle construction and bridge, kraal and house building. Though it was estimated in 1880 that 166 000 acres (67 000 hectares) of indigenous forest existed in Natal, the nature of these forests, their inaccessibility and the fact that the best timber had already been cut out necessitated the importing of timber from abroad. In the 1870s imports totalling 2,25 million cubic feet (64 000 cubic metres); in the 1880s, 8,75 million cubic feet (248 000 cubic metres); and in the 1890s, 26 million cubic feet (736 000 cubic metres).[65]

In an attempt to reduce this enormous import bill the colonial authorities appointed conservators of forests to regulate cutting in some forests. The lieutenant governor, Sir Henry Bulwer, like so many colonial officials was personally interested in botanical matters and in 1878 he appointed an official commission to look at the problem of deforestation. Its report in March 1880 recommended the establishment of a forestry department and of state subsidised afforestation. Forests on crown land were closed to sawyers until January 1884 in an attempt to help growth recover. Though the report was shelved because funds were not available to implement it, the authorities and especially the surveyor-general, P.C. Sutherland, actively encouraged the planting of Australian exotics. These had been grown in Natal as early as 1839 but it was only in the late 1860s that widespread private afforestation schemes were undertaken. The principal species planted were blue gums (*Eucalyptus globulus*) and black wattle (*Acacia mearnsii*). The latter, which was introduced in the mid-1860s, became of major importance to the economy of Natal. By the year 1904 there were over 75 000 acres (30 000 hectares) of wattle plantation in the colony.[66]

In the later Victorian era the major tree planters in Natal were messrs Baynes, Blackburrow, Corduke, Henderson, Sutton, Topham and Phillips. These individuals were supplied with seedlings by various sources: local and Cape nurserymen, direct from Australia and, though this is less well known, from Kew. Kew supplied Natal with seed and plants of both coniferous species and of many kinds of fruit tree. Indeed in the late 1860s one of the principal Kew exports to Natal was fruit trees, which were sent to the Durban botanic gardens and thence to the Natal midlands. By 1881 one correspondent informed Kew that 'there will soon be a thousand orchards in Natal and there are a hundred English pheasants near Durban'.[67]

Robert Topham was closely involved in importing timber trees from Kew. He told Hooker he had *Eucalyptus* which had grown 60 feet (18 metres) from seed in five years. Topham, with Kew's assistance, successfully diversified into growing other genera such as *Pinus radiata*, *Araucaria heterophylla*, (Norfolk Island pine), cedars and English and Turkey oaks. In 1874 he recorded the new enthusiasm of colonists for tree planting:

The taste for planting is making good progress. I have known people take trees in open boxes 400-500 miles in ox wagons and no one would do that except for the love of them.[68]

All this planting altered the Natal landscape. As the indigenous escarpment forests receded, plantation forests replaced them. In addition most farms planted windbelts and isolated clumps of Australian trees to provide shelter for stock and cover for game. Though the fate of the indigenous forests was genuinely lamented by the botanically minded settlers, the new

---

65 Donal P. McCracken, 'The indigenous forests of colonial Natal and Zululand', *Natalia*, 16, (1986).

66 For contemporary comment on wattle growing, see T.R. Sim, *Tree planting in Natal*, (Pietermaritzburg, 1907), pp.125-148.

67 K.A.: signature illegible, 1 July 1881, written from Algoa Bay; South African letter vol 189).

68 K.A.: Topham, 15 September 1874.

plantations were heralded as a considerable achievement which greatly enhanced the colony. In 1883 Topham wrote to Sir Joseph Hooker of the contrast between 'the brown wind-blown sunburnt hell of fifteen years ago and the green cool damp sweet shade of the present'.[69]

The crisis in Natal botany

THE LATE 1870S AND EARLY 1880s were difficult years for Natal. The very existence of the colony was threatened by the outbreak firstly of the Anglo-Zulu war on its north-east frontier in 1879 and two years later of the first Anglo-Boer war on the northern frontier. This later conflict resulted in Kew receiving a collection of lichens from Major H.W. Fielden of Wells-next-the-sea in Norfolk, who collected when camped on the slopes of the Drakensberg 'waiting for the settlement of the Boer difficulty'.[70] The Anglo-Boer war brought to Kew fewer dramatic tales in the colonists' letters than had the Anglo-Zulu war, during which, for instance, Medley Wood had written Sir Joseph Hooker a dramatic account of how he had had to flee his home in Inanda after the disastrous British military defeat at Isandhlwana, 'as my place lies directly in the road a Zulu army would take on their way to

A note dated 1870 from Robert Topham to Sir Joseph Hooker. Topham was from a slightly different mold from most of the colony's plant collectors in that he was a businessman. He was especially interested in colonial afforestation. (Courtesy: Royal Botanic Gardens, Kew)

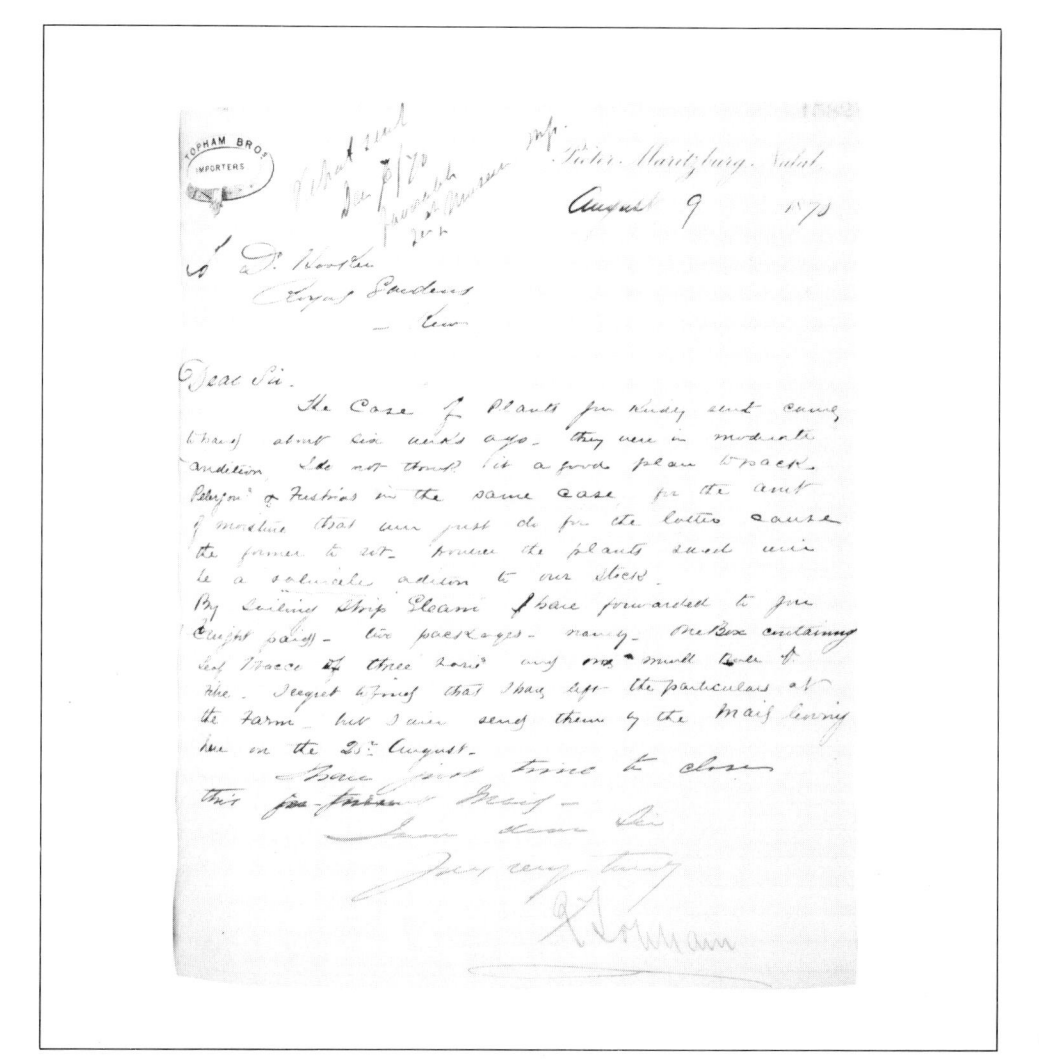

69 K.A.: Topham, September 1883.

70 K.A.: H.W. Fielden, 18 April 1881 and 24 October 1882.

Durban'. Though Wood was forced to 'leave everything behind' when he and his wife set out on their 17-hour epic trek to Verulam, he could not resist taking with him two parcels of dried specimens for Kew.[71]

While Colonel Bowker enjoyed the Anglo-Zulu war immensely, and sent Kew a mounted hoof from an artillery horse killed at Isandhlwana,[72] the war served to exacerbate an already bad situation for Keit. It had been difficult enough for him to recruit his African labourers and for a while he had been reduced to using convict labour and, to Kew's horror, carrying out manual labour himself, but when the war broke out his hard-found recruits disappeared. A drought with only 28 inches (71 centimetres) of rain falling in 1878 instead of the average 40 inches (102 centimetres) had meant that all efforts had had to be concentrated on carrying buckets of water to plants from the ponds and the tank. White ants attacked the greenhouse, the summer houses, the seats and plant labels and within a relatively short time the garden looked dilapidated and parched. Strangely at the same time Britain was experiencing a prolonged period of wet weather which resulted in an agricultural depression. In 1879 a very severe hailstorm resulted in about 40 000 panes of glass being smashed in Kew's conservatories.[73]

Keit further suffered because he was unable to leave the gardens and collect new plants for them and for exchange with overseas botanical institutions. This was because, for financial reasons, the horticultural society had not provided him with an assistant gardener who could take the twice-daily meteorological readings in the weather station established in the grounds in 1873. During his time as curator Keit only twice escaped to collect and on both occasions the weather marred his expeditions: in 1874 it was too wet and in 1877 it was too dry and hot. In 1875 he told Joseph Hooker, 'beyond our gardens I see and hear little'.[74]

Personal problems also affected Keit. In September 1874 he had married Anna Louise Currie, the daughter of a prominent Durban citizen. They lost one child from dysentery and another nearly died of the same disease. In 1880 Keit told Thiselton-Dyer, 'Doctor and undertaker have shared between them the small savings I had managed to scrape together'.[75]

But most dangerous for Keit's position was the hostility he met from a section of influential townsfolk. He was accused of not doing his job properly and of being distant towards visitors. Kew recognised the true nature of his difficulties early on: there was no money to develop the gardens; it was too big to manage on the resources available; too far from town so few citizens bothered to visit; those who were interested in botany preferred to pursue their interests individually rather than through the horticultural society; Keit was reserved and at times awkward in handling difficult people; and lastly he was a German in a very English colony. In 1874 Joseph Hooker offered to help Keit get another post outside Natal. But by then Keit was married and a sense of pride held him in his position. He did, however, plead, 'Should I fail then I trust you will not forsake me'.[76]

In 1880 matters began to come to a head. Kew was not prepared to have its man 'cast to the wind' if it could at all be helped. In the annual report for Kew, Hooker noted of the Durban gardens:

The state of decay and entire lack of local encouragement... is quite distressing. The superintendent seems... to carry on his work under great disadvantage.[77]

---

71 K.A.: Wood, 25 April 1879.
72 K.A.: Bowker, 4 May 1880.
73 *Kew Report*, (1879); and K.A.: Keit, 2 May 1873 and 23 September 1873.
74 K.A.: Keit, 8 April 1875.
75 K.A.: Keit, 15 July 1880.
76 K.A.: Keit, 6 December 1874.
77 *Kew Report*, (1880).

Thiselton-Dyer took up Keit's cause in another quarter. Keit had told him how his annual report had neither been printed nor laid before the legislative council as was normal. Thiselton-Dyer wrote to the colonial office:

It is possible in Natal that these documents are regarded as departmental papers of little moment. I am, however, to point out that they are extremely useful to such an institution as this in as much as they constitute an official record of the success or failure which attends the introduction of new economic plants and give information on such subjects which is often highly serviceable to similar institutions in other colonies.[78]

The report, containing Keit's defence, was duly published. It was, however, not enough to save him. A threat by the colonial government to withdraw the gardens' grant prompted the horticultural society to call a public meeting to solve the problem. 'I do not expect Mr Keit will suit us to continue as curator. He would make a very excellent second,' wrote James Aiken, the society's treasurer, to Sir Joseph Hooker. On 3 February 1882 it was resolved at the meeting to abolish the old Natal Agricultural and Horticultural Society and found a new society to run the gardens. It was also agreed to find a new curator.

Two contenders vied for the posi-

'Durban from Mr Currie's residence, Berea 1873' by the botanical artist Thomas Baines. Henry William Currie, a later mayor of Durban, was to be the father-in-law of William Keit the recently appointed curator of the nearby Durban botanic gardens. The plants in the small flower bed have been carefully labelled.
(Courtesy: Durban Local History Museum)

78 Miscellaneous reports: Natal botanic gardens, 1866-1916, vol 311, Thiselton-Dyer to the colonial office.

John Medley Wood (1827-1915)
Wood dominated Natal botany from the 1880s until his death. He was curator and then director of Durban botanic gardens from 1882 to 1913. He also established the colonial herbarium in 1885. He was a prolific plant collector and published many treatises on Natal's botany. In 1912 Kew honoured 'the father of Natal botany' by dedicating a volume of *Curtis's Botanical Magazine* to him. In 1913 the University of the Cape of Good Hope awarded him an honorary doctorate. His most famous work is the six volume *Natal Plants* which appeared between 1898 and 1912. (Courtesy: National Botanical Institute, Pretoria)

tion, the self-opiniated J.L. Meade, who considered himself better qualified to be curator than anyone inside the colony or abroad, and a rather embarrassed John Medley Wood.[79] The latter was duly appointed and a new Durban Botanic Society established to run the gardens. To the colonial secretary Aiken wrote, 'This appointment will be warmly approved not only, as it is, in the colony, but by Sir Joseph Hooker of Kew'. At the same time Wood rather pompously informed Kew that his private correspondence would cease and he would now correspond and exchange with Kew officially as curator but also asked Hooker for 'hints' on how to run a botanic gardens as he had never had experience in one before.[80]

William Keit set up his own private nursery and was soon appointed curator of the town's parks. He prospered and did much to beautify Durban. The affair of his dismissal would have rested with Wood's appointment had Sir Joseph Hooker not been extremely disturbed by its implications. The untimely death of John Sanderson in March 1881 had taken from the colony one of the few men who favoured botanical advancement through the botanic gardens. On Hooker's instructions Thiselton-Dyer wrote to R.H. Meade of the colonial office in London. He pointed out the unsatisfactory nature of botanical enterprise in Natal, especially relating to the lack of public support for botanical establishments. Meade showed the letter to the colonial secretary, Lord Kimberley, who in turn forwarded it to the governor of Natal, Sir Henry Bulwer, with the comment, 'Sir Joseph Hooker's remarks appear to me deserving of careful consideration'.

Twelve days later, on 27 November 1883, Aiken was writing to Bulwer justifying his action and stating that Sir Joseph Hooker had been kept fully informed of developments. Bulwer had already initiated the forestry report mentioned earlier and now he collected evidence on general botanical matters. These he had published in 1884 under the title, *Correspondence and reports relative to the state of the botanical enterprise in Natal*.[81]

Hooker's direct intervention in Natal's botanical affairs was to have profound and beneficial repercussions. Wood, who had had the sense to support Keit in a letter to Hooker, found himself in a strong position.[82] Before the end of 1882 the new botanic society had brought out from Britain a Kew-trained gardener to assist Wood. They also paid £13 for the erection of a fernery. The following year a hard surface was put on Botanic Gardens Road, a sandy track which led along the Berea to the gardens and which was very difficult for carriages to negotiate. By the end of 1883 Wood had cut down many of the old unsightly trees and had published the gardens' first guide. It was entitled *A guide to the Natal Botanic Gardens, containing plan of the gardens, byelaws, etc.,*

---

79 K.A.: Colonial secretary's office, vol 844, no 664.

80 K.A.: Wood, 1 March 1882.

81 A copy of this report is housed in the Natal archives.

82 K.A.: Wood, 8 May 1882.

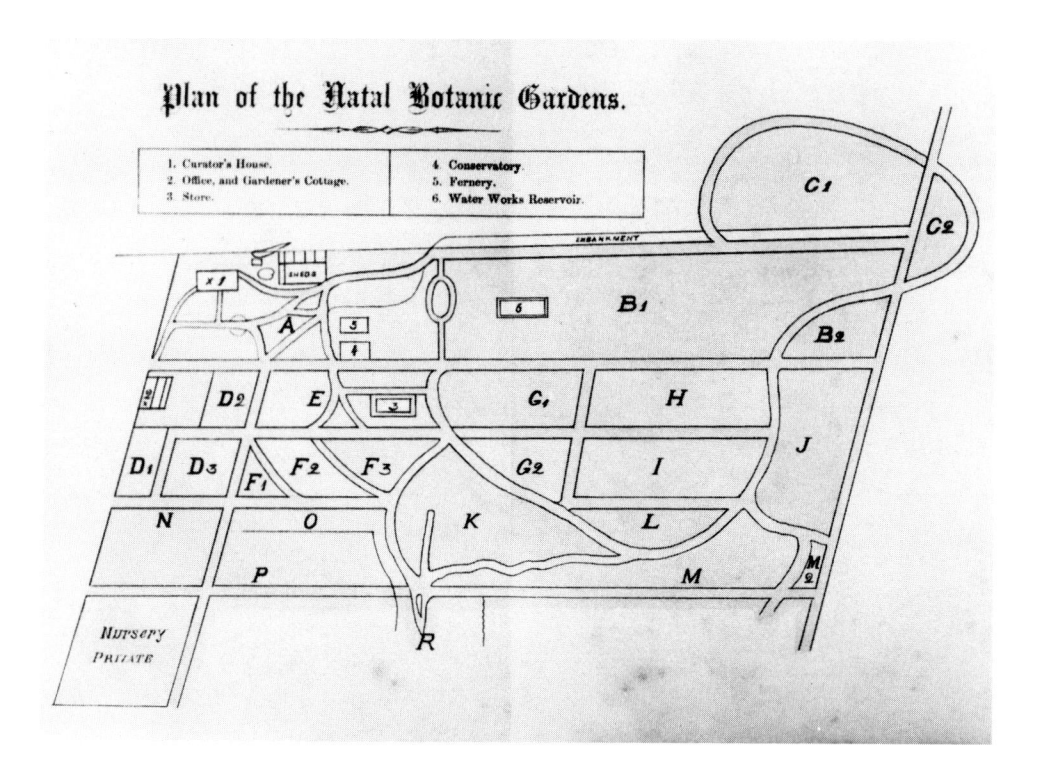

Plan of the Durban botanic gardens which featured in the garden's first catalogue of plants, which was compiled by Medley Wood. This contained 56 pages and was published in 1883. The paths in the upper part of the gardens remain largely unchanged to this day.

Carriages making the journey from town in the 1870s brought visitors to St Thomas' Road entrance to the Durban botanic gardens. (Courtesy: Durban Local History Museum)

and catalogue of plants and it was 56 pages long. That year also saw the passing of law number 21, 'To incorporate the Durban Botanic Society', which permitted the society to borrow money. With this backing and an annual rainfall of 44 inches (112 centimetres) Wood was soon, with the assistance and support of Kew, to direct one of the most attractive and successful botanic gardens in the British empire.

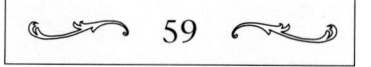

Chapter 5

The Indian summer

The new Kew regime

As the mid-1860s were years of change for Kew and for Natal botany, so the middle years of the 1880s divide one era from another. For Kew the major development was the retirement of Sir Joseph Hooker as director in 1885. Two years previously, 26 years after its commencement, he had completed *Genera Plantarum*. George Bentham, the co-author, died in September 1884.[1] Hooker considered his own health to be poor and retired to his country house, 'the camp', in November 1885. In fact he was to live for another 26 years, dying only in 1911 at the age of 94. He devoted much energy, ably assisted by D.D. Jackson, to preparing the multivolumed *Index Kewensis*, a list of those seed-bearing plants botanically described between 1753 and 1885. The first part of this work appeared in 1895. Hooker also completed his flora of British India during his years of retirement. He was honoured by being created a knight grand commander of the star of India in 1897.

Joseph Hooker's successor was his son-in-law, William Thiselton-Dyer, who had served an apprenticeship as assistant director for the previous decade. Thiselton-Dyer was an imperious man and, though highly competent, was much less charming than the Hookers. He was a disciplinarian who was inclined to run Kew along military or strict monastic lines: on one occasion he referred to himself as the 'botanical pope'.[2] He hyphenated his name in 1891 and in 1899 he was knighted.

The new assistant director of Kew was Daniel Morris, a gold medallist from Trinity College, Dublin, who had served in botanical institutions in Ceylon and Jamaica. While Thiselton-Dyer dealt with much of the correspondence with colonial botanic gardens and stations, most of the routine correspondence concerning plant identification rested with Morris and the assistant curator in charge of tropical plants, William Watson. Morris became increasingly involved in studying the flora of the warmer regions of the world: so much so that when in May 1896 Beatrix Potter requested some help, he proclaimed, "I am exclusively tropical."[3]

Professor Oliver remained keeper of the herbarium until his retirement in 1890 but, as in the past, he left most of the correspondence to the director and the assistant director. Oliver's successor, John Gilbert Baker, did, however, maintain direct contact with overseas institutions but by then the Kew scientific establishment had become larger and the volume of work proportionately greater. Other

---

1 Turrill, *Kew Past and Present*, p. 34.

2 Blunt, *In for a penny*, p. 190.

3 Leslie Linder (ed) *The Journal of Beatrix Potter*, (London, 1966), pp. 413-5, 424, 426 and 429.

Sir William Thiselton-Dyer (1843-1928) 'the botanical pope' Thiselton-Dyer was director of Kew from 1885-1905. His marriage to the daughter of Sir Joseph Hooker ensured the dynasty at Kew which lasted in all for 64 years. Less of a diplomat than the two Hookers, none the less Dyer brought Kew fully into the imperial orbit. He was a keen advocate of the establishment of colonial botanic stations. In 1887 he started the *Bulletin of miscellaneous information*, a periodical which featured Natal on several occasions.
(From, *Kew Guild*, 1893)

botanists working in the herbarium in this late Victorian period included messrs Brown, Hemseley, Massee, Rolfe, Stapf and Wright. The scale of their workload can be judged by the fact that by the Edwardian period, as well as carrying out their normal duties, the herbarium staff were identifying in the region of 3 000 specimens annually for members of the public.

FOR KEW'S RELATIONSHIP with the empire, Thiselton-Dyer's accession to the directorship heralded the cementing of a special bond between Kew, the colonial office and the Indian office concerning botanical matters. It is significant that the arch-imperialist colonial secretary, Joseph Chamberlain, took special interest in promoting Kew and indeed was largely responsible for the government financing the south wing of the temperate house in the late 1890s.

Thiselton-Dyer was particularly associated with the colonial office which in February 1886 formulated a strategy for the establishment of minor botanic stations in the West Indies. The scheme was taken as a

HELIOPHILA SCANDENS (Bridal wreath) This member of the Brassicaceae family likes to grow in shady spots among shrubs. It was first described by Harvey. A specimen was sent to Kew by Medley Wood in 1885. (*Curtis's Botanical Magazine*, T7668, [1899])

4 Donal P. Mc Cracken, 'The development of Botanic institutions in nineteenth century Natal and west Africa', *Journal of the University of Durban-Westville*, new series 1, (1984), p. 110.

5 K.A.: Bowker, 21 April 1885; and Wood, 30 June 1885.

6 *Kew Bulletin*, no. 9, (September, 1887), pp.12-3.

7 Notes on aspects of Natal's botany may be found in the *Kew Bulletin* for the years 1887, 1888, 1890, 1893, 1895, 1898 and 1900.

8 K.A.: Wood, 8 May 1882 and 9 November 1894; and *Kew Bulletin* no. 29, (May 1889), pp. 122-6.

prototype for the west African botanic stations established in the late 1880s and early 1890s.[4] To ease the work burden for his hard-pressed staff, Thiselton-Dyer encouraged colonial botanic gardens and stations to exchange plants of possible economic value with each other directly, though Kew continued to mastermind the introduction of the valuable species to colonial settlements.

Thiselton-Dyer was also an active participant in the setting up of the great colonial and Indian exhibition in London in 1886. This exhibition created many problems for the smaller, poorer colonies like Natal. Much to Medley Wood's disgust, the Durban Botanic Society asked Kew to lend the Natal stand at the exhibition a number of Natal plants growing at Kew.[5] Natal produced an excellent display which included yarn spun from the fibre of Natal plants, old stone implements, a hundred botanical paintings of Natal plants by Sam Large which were later bought by Kew, wild tea *(Geranium incanum)*, both concrete and crystallised sugar, linseed, extract of American aloe, tobacco both cut and leaf, groundnut oil cake, seeds of fenugreek and the tanning agents used in Natal and extracted from *Elephantorrhiza elephantina (E. burchelli)* and *Harpephyllum caffrum*, as well as numerous ferns and flowering plants. Kew, which received most of the botanical exhibits when the exhibition closed, was especially proud of the high-quality Natal tea on show as they claimed to have provided the original plants to James Brickhill, who won a gold medal at the exhibition for his tea.[6]

SUCH EVENTS AS THE 1886 exhibition were mentioned in a new monthly journal published by Kew from 1887. Fully titled *Royal Botanic Gardens, Kew: Bulletin of miscellaneous information*, it was usually referred to simply as the *Kew Bulletin*, though its name was not officially changed to this until 1946. In the 1880s and 1890s, the *Kew Bulletin* carried notes or articles on Natal products such as aloes, rubber, dried plants, forestry, fruits, lily culture, maize production, museum specimens, tea cultivations and unera fibre, as well as numerous references to indigenous plants received at Kew.[7]

In 1882 Medley Wood had suggested to Kew that it would be a good idea to publish a list of botanic gardens and their curators. This was done periodically in the *Kew Bulletin*, the first occasion being in 1889 when botanic gardens in Africa corresponding with Kew were listed as Cape Town, Graaff-Reinet, Grahamstown, King William's Town, Port Elizabeth, Uitenhage, Durban, Pietermaritzburg and Asaba in the Niger territories.[8]

In addition to the *Kew Bulletin*, *Curtis's Botanical Magazine* continued to be published by Kew and contained more and more references to Natal plants: in 1885, for example, the magazine described *Dioscoria crinita, Eucomis bicolor, Calanthe natalensis* and *Aloe bainesii*.⁹

Of much more importance, however, for South African botany was the decision by Kew to revive the moribund *Flora Capensis*. Although the preface to volume III of the Flora published in 1865, had made mention of a successor volume 'shortly to be in preparation for the press', practically no material relating to it was found among Harvey's papers after his death in 1866. Though Harvey's 'coadjutor', Sonder, did not die until 1881, he took no further interest in the project. As early as September 1866 Sanderson was inquiring of Kew who was to carry on with the much-needed flora.¹⁰ Despite agitation from Sir Henry Barkly, governor of the Cape from 1870 to 1877, Kew felt that it could not proceed with the undertaking until finance was provided by the governments of the South African colonies. Though he had already done much to advance botany in the West Indies, the lieutenant governor of Natal, Robert Keate, could not recommend a grant from the impoverished coffers of the colony. Nevertheless pressure from prominent colonists resulted in £75 being granted at the end of 1872, and the *Kew Report* announced that Thiselton-Dyer would undertake this work.¹¹

By 1874 little had been done, though over the next two years Thiselton-Dyer published three papers on South African succulents and worked on the hitherto much neglected subject of South African aloes. By the end of that period Thiselton-Dyer, in his new position as assistant director, was carrying much of the administrative work of Kew. Inevitably the project was postponed; by 1884 the former Cape colonial botanist, Rev. John Brown, was even talking of undertaking his own government-financed flora of Natal, though nothing came of this.¹² It was not until the 1890s that Thiselton-Dyer, once again under pressure from Barkly, undertook the supervision of the *Flora Capensis*. This time his efforts were successful and volumes began to appear from 1896. The project was completed in 1933, only four years before the flora of tropical Africa was finished. Both the Cape and Natal governments sponsored Thiselton-Dyer, both administrations receiving 25 free copies of the work. In the case of Natal it was the governor, Sir Walter Hely-Hutchinson, a very keen botanist, who had been able to persuade the self-governing administration to include a grant for the flora in the estimates.¹³ The 30-year hiatus in the production of the *Flora Capensis*, while serious, did mean that the new volumes were able to include more recently discovered plants. It also ensured that South Africa was kept to the fore at Kew in the post-Victorian period.

EUCOMIS BICOLOR (Purple pineapple flower) This member of the Liliaceae family was discovered by Christopher Mudd in Natal in the mid-1870s. He sent it to Messrs Veitch in Britain in whose nurseries it flowered in 1878. By 1884 it was seen to be growing outside the Great Palm House at Kew. *(Curtis's Botanical Magazine,* T6816, [1885])

9 *Curtis's Botanical Magazine,* (1885), t6804, t6816, t6844 and t6848.

10 K.A.: Sanderson, 20 September 1866.

11 K.A.: Sanderson, 20 March 1872, 1 July 1872, 11 December 1872, 19 December 1872 and 2 March 1874; and *Kew Report,* (1872).

12 K.A.: A.H. Mitchell, 21 October 1884.

13 See K.A.: Hely-Hutchinson, 15 March 1897 and 21 January 1898.

63

THE CLOSING DECADES OF the nineteenth century were extremely exciting ones for scientific development at Kew. Less dramatic were developments in the gardens proper. The pioneering stage of laying out had largely passed. There were some more additions: between 1887 and 1899 an alpine house, refreshment pavilion, filmy fern house and two new wings of the temperate house were built. Kew also acquired the Queen's Cottage; and a bamboo and lily garden were established.[14] In 1886, a year after Sir Joseph Hooker's retirement, the curator, John Smith II, bowed out and was succeeded by George Nicholson, who was to be the curator in charge of the first women gardeners, appointed in 1896. With Hooker and Smith gone, a stricter regime now prevailed in the gardens. Thiselton-Dyer was not noted for his diplomacy in dealing with gardeners; and at the turn of the century the foreman of the arboretum, William Dallimore, was nicknamed De Wet by the apprentices because they never knew when he was going to sneak up on them.[15]

The attraction of Kew in the 1880s and 1890s was not so much innovative garden architecture but rather that the grounds were now well established: the vistas were flanked by large trees, the shrubs big enough to give dramatic floral displays and the conservatories packed with large and extraordinary exotic vegetation. Never before had Kew been so popular with the general public. Yet this popularity created problems for the institution. The large number of visitors caused much damage to the gardens, despite the vigilance of Thiselton-Dyer's famous uniformed Kew constabulary of whom he was officer-in-charge.

It was not just the domestic public who trespassed on Kew's goodwill: overseas correspondents used the institution as an agent to forward their mail to addressees in Britain or on to other colonies, a practice which had earlier been used by McKen to send letters to his brother in Newfoundland. Kew was also often asked to raise seeds sent by a colonist and then to forward the grown plants to the colonist's friends in Britain. Captain Garden, W.P. Robertson and John Sanderson of Natal all resorted to this annoying practice. But one of the most time-consuming tasks for senior Kew staff was entertaining and showing round the gardens either important British visitors or eager colonists on a visit home. In the 1850s Joseph Hooker had written to his grandmother:

My mother and sister will tell you that of the hundreds of aristocrats who detain my father at the Garden for hours waiting their arrival and then drag him through every house and acre, there are not half a dozen who... have shown the slightest politeness in return.[16]

Colonists treating Kew as one of the tourist attractions they had to include on their itinerary could be equally annoying, often complaining loudly if they found the director not at home when they called. Some correspondents wrote in advance requesting special favours. John Sanderson when in Britain wanted to be present at Kew when his wardian case was opened so that the ladies in his party could see the event. In 1880 Dr Sutherland wrote to Kew of a Natal colonist visiting Britain:

Mr Beningfield is one of our leading men, and though he takes no interest in Botany himself, Mrs Beningfield, I have no doubt, will be pleased with any little attention you can show her.[17]

It is to Kew's credit that it bore these many importunities with good grace. Of course there were many colonial visitors whom it was genuinely

---

14 See Bean, Chapter IX.

15 Blunt, *In for a penny*, p.194.

16 Allan, p.157.

17 K.A.: Sutherland, 5 April 1880. See also D.Arnot, 15 August 1876 and 24 August 1876; Sanderson, 11 June 1861; and K. Saunders, 26 April 1896.

pleased to entertain and whom it fired with further enthusiasm to promote botanical developments in their respective colonies. Sutherland recalled once in a letter to Sir Joseph Hooker how several visits to Kew when he had met eminent botanists had proved to be stepping stones in his life.[18] Kew was also eager to welcome those colonists who went to the considerable inconvenience of accompanying a wardian case on the voyage to Britain and then from the docks to Kew. Indeed it was the attitude of such enthusiastic amateurs that did much to forge the links of the great imperial chain of botanical cooperation between Britain and her colonies, and between the different colonies themselves, an attitude that was picturesquely summed up by Robert Jameson of the Durban Botanic Society, and discoverer of the Barberton daisy *(Gerbera jamesonii)*, when writing to Kew in 1892: 'I shall feel privileged in becoming the "slave of your lamp".[19]

Natal in the new era: The Pietermaritzburg botanic gardens

BY THE MID-1880s NATAL was beginning to recover from the wars and drought which had recently plagued the colony. Agitation for self-government along the lines granted in 1872 to the Cape bore fruit in 1893 when responsible government was granted to Natal. In the field of botany a new order was also emerging. William Keit was no longer at the Durban botanic gardens and John Sanderson was dead. The Pietermaritzburg botanic gardens, after surviving a fraud scandal, flourished as a producer and distributor of seedlings under a succession of nondescript curators, one of whom claimed in 1885 that he didn't 'know anything about herbariums.'[20] The botanist R.W. Adlam might well have developed the gardens along more scientific lines and indeed for a short period he corresponded with Kew over fern spores which he wanted, but in 1889 he quarrelled with the society's council and resigned.[21] That month the *Gardeners' Chronicle* dryly commented that the Pietermaritzburg botanic gardens looked 'like St Edmund's Abbey when Abbot Sampson was installed (*vide*, Carlyle).'

The Pietermaritzburg botanic society, much to Medley Wood's disgust, advertised for a 'practical gardener' rather than a curator and offered a salary of £12 per month.[22] They were fortunate to obtain for this post the competent G. Mitchell, who remained at the gardens until his death in 1900.[23] He thinned out the existing stands of timber in the gardens, laid out paths and lawns and generally created an attractive park.

When the Durban botanic society applied for an increased grant in 1890, its Pietermaritzburg counterpart did likewise which greatly angered Hooker. The result was that the Natal

RICHARDIA REHMANII This member of the family Araceae is now known as *Zantedeschia rehmanii*. It is found on rocky outcrops, forest margins and damp ground. In 1888 it was sent to England by the botanist R.W. Adam a year before his short period as curator of Pietermaritzburg botanic gardens. Kew received its specimen from Medley Wood in 1893. It was described by N.E. Brown and named in honour of the Polish botanist Anton Rehmann who visited South Africa twice in the 1870s. *Curtis's Botanical Magazine*, T7436, [1896])

18 K.A.: Sutherland, 8 August 1883.

19 K.A.: Jameson, c June 1892.

20 K.A.: Wood, 24 July 1885.

21 K.A.: Adlam, 27 April 1889.

22 K.A.: Wood, 24 July 1889.

23 See *Natal Witness*, 19 November 1900.

The lake with a boat in Pietermaritzburg botanic gardens In 1890 the gardens offered this pond to the colonial government for 'the promotion of pisiculture'. The offer was declined so instead the lake was stocked with swans, which proved a great attraction to the public and to 'a wild animal' which in 1898 ate three of them.
(Courtesy: Killie Campbell Africana Library)

24 K.A.: Hely-Hutchinson, 27 October 1893 and 27 March 1895.

25 In letters to Kew, Wood had a tendency to refer to Wylie merely as 'the gardener' : see, for example, K.A.: Wood, 31 March 1885, 13 November 1888 and 27 February 1889. However, see also Wood's tribute to Wylie: K.A.: Wood 27 August 1897. For details of a row over the role of Frieda Lauth in the preparation of *Natal Plants*, see K.C.A.L.: Durban Botanic Society, Book 5, KCM 43069.

government set up a botanic commission under Sir Theophilus Shepstone to examine the possibility of bringing the two gardens under one government department and also the possibility of establishing a third gardens in the upland region of northern Natal. Partly because of the death of Shepstone, but also because of what the governor referred to as 'the absurd jealousy existing between the two towns', nothing came of either proposal.[24]

The result for the Pietermaritzburg gardens was that it increasingly served as a public park and a commercial nursery, much to the annoyance of local nurserymen. Only in the Edwardian period did Kew have much contact with the gardens; from 1904 it provided the botanic society with a succession of curators, including Alexander Hislop, W.E. Marriott and W.J. Newberry. Occasionally Kew was in correspondence with the Pietermaritzburg gardens but most of the developments in the early years of the new century were encouraged by annual visits from Medley Wood.

The Durban botanic gardens

AS THE NINETEENTH CENTURY drew to a close, enthusiasm for botanical activity increasingly concentrated on the Durban botanic gardens which, under Wood's curatorship, was entering its heyday. Though Wood tended on occasions to be rather self-opinionated and to play down the contribution of others employed at the gardens,[25] none the less he proved the ideal man for the job. He was a practical botanist, a keen collector and exchanger of plants and seeds, and a dedicated correspondent with other botanists and botanical institutions. In the 31 years that Wood controlled the gardens he was able to establish one of the most beautiful botanic gardens in the British empire.

Wood was aided in his undertakings by a variety of new factors. The

Sir Theophilus Shepstone in his garden in Natal. Shepstone was secretary for native affairs in Natal and a man of considerable influence with the colonists and the Zulus. He was interested in plants and supported the colony's two botanic gardens when possible. He also used this garden to acclimatise fruit trees introduced in 1870.
(From, *The Gardeners' Chronicle,* 18 April 1887)

water problem was solved by the drought breaking and by the construction of a 50 000 gallon (227 305 litre) municipal reservoir in the gardens. The botanic society was now legally able to raise a loan which enabled the construction of a new office, a gardener's house and a new conservatory which cost £287. The gardener's house was occupied by James Wylie, the newly arrived Kew man. The decision of the botanic society to pay for a gardener was the key factor in the gardens' progress. Wylie's appointment meant that there was now a trained man to keep a permanent eye on the gardens, thus allowing Wood to come and go as he pleased. He was accordingly often away collecting for lengthy periods.

In 1889 a new curator's house was built in the gardens, in 1894 a telephone link with the town was established and in 1897 a large jubilee conservatory was built at a cost of nearly £3 000. It was opened by the governor, Hely-Hutchinson, on 8 August 1898. Interestingly Thiselton-Dyer told Wood that such a structure in Durban, would be nothing but a 'nuisance'. Wood justified the construction by claiming that 'the public will have it.'[26] Kew was much more interested in the fine collection of African, Asian, Australian and south American plants listed in the catalogues which Wood produced in 1889 and 1897, and found the work he was doing in the new colonial herbarium especially interesting.

The colonial herbarium

WHEN MEDLEY WOOD BECAME curator in 1882 he immediately established an herbarium in an old wood and iron hut in the gardens. He had two cabinets for specimens. As a basis for the collection he had his own dried plants from Inanda, a collection of Rev. Buchanan's Australian and central African specimens, about 2 000 specimens in poor condition which already belonged to the gardens and had been collected by Gerrard and McKen and

26 K.A.: Wood 11, November 1898. In 1905 a fernery costing £900 was constructed beside the jubilee conservatory.

67

Interior of the jubilee conservatory in the Durban botanic gardens. This conservatory was finally dismantled in the 1940s when it had deteriorated beyond repair.
(Courtesy: Durban Local History Museum)

The colonial herbarium of the Durban botanic gardens was built in 1902. By 1913 it contained 45 000 dried specimens nearly a third of which were South African plants. In the background of this photograph is the attractive curator's house which was erected in 1890.
(Courtesy: Durban Local History Museum)

27   B.D. Schrire, 'Centenary of the Natal Herbarium, 1882-1982' *Bothalia*, 14,2, (1983), pp. 223-36; and K.A.: Sutherland, 28 May 1883; and Wood, 19 May 1885.

finally a large parcel of specimens sent by Kew to Wood.

Kew pressed strongly for this herbarium to become an official colonial institution. The issue was successfully taken up by Wood, P.C. Sutherland and Sir Benjamin Greenacre and, in 1885, two acres (0,8 hectares) of the Durban botanic gardens were ceded by the botanic society to the Natal government for the establishment of a colonial herbarium.[27] Wood was appointed director and an initial government grant of £60 was given to the new institution. In that year the governor also presented the new institution with a microscope.

In 1887 the herbarium had 5 000 specimens and by 1901 it contained 27 000, a third of which were South African specimens. This herbarium, though rarely visited by the general public, was to serve the colony well. Wood used it both as a herbarium and as a laboratory, in which he experimented with fibres and carried on

A note dated 1895 from the curator of Durban botanic gardens, John Medley Wood, to Sir William Thiselton-Dyer. It gives details of some of the specimens he was despatching to Kew. Wood was indefatigable both as plant hunter and correspondent with Kew from the 1870s until his death in 1915.
(Courtesy: Royal Botanic Gardens, Kew)

research into numerous problems of plant disease and adaptation of exotics to Natal's climate.[28] In 1884, for example, he had worked on the fungus which had blighted the coffee plantations, discovering a new variety of *Hemiteia* which was named *woodii* in his honour. Despite a return of coffee disease in 1893, planters continued to apply to Kew for seed. As with many similar demands Kew usually referred such correspondence back to the Durban botanic gardens.

Wood also advised planters about fungal attack on bananas in Natal in 1887 and on various sugar cane diseases. Though cane smut in China cane had greatly diminished by the mid-1890s, Wood encouraged the growing instead of the Uba variety. He claimed to have discovered it in an old warehouse at Durban docks and had grown it in the gardens before distributing it widely. Wood believed the education of farmers was vitally important for the colony. He was particularly concerned following a visit into the interior when he discovered settlers who did not know the difference between a *Eucalyptus* and an *Acacia*, let alone realise the large number of varieties of *Eucalyptus* growing in the region. To Kew he wrote that such ignorance 'shows very

28 K.A.: Wood 23 August 1887, 17 October 1887, 23 November 1887, 15 June 1888; and Kew memorandum, 29 November 1887. See also *Kew Bulletin*, (1888), pp.84-5.

KNIPHOFIA MODESTA This member of the family Liliaceae is now known as *K. parviflora*. It was first discovered in 1884 by William Tyson in the mountainous country of East Griqualand, which today is part of Natal. Medley Wood later found it in the colony itself. (*Curtis's Botanical Magazine*, T7293 [1893])

plainly the value of a Colonial Herbarium.'[29]

The herbarium was used for identification and also naming of new species. Wood even named species sent to him by Colonel 'Butterfly' Bowker on which he had found butterflies feeding.[30] For MacOwan's 'Herbarium Normale Austro-Africanum', Wood sent duplicates from the colonial herbarium to Cape Town for distribution to herbaria overseas. He also carried out research on plant dyes, aloes, the medicinal properties of plants, and poisonous plants, this last involving him in the investigation into a murder case in 1898.[31]

The herbarium soon became an important seed bank (for the colony). In this venture Kew was very accommodating and in regularly sending the Durban botanic gardens bags of seeds from the Kew seed bank. Wood was thus able to provide colonists with such crop seed as indigo, teff, manihot rubber, sunflower and pineapple, as well as the more established varieties.

In addition to these many activities, Wood was an 'indefatigable correspondent' with Kew and other botanical institutions. He was also the author of a number of plant lists and catalogues and of an annual report which was sometimes as long as 80 pages. His most famous undertaking was his six-volume *Natal Plants* which appeared between June 1898 and April 1912. Both Plant and McKen would have liked to have undertaken such a flora but it was only in 1894 that Wood made a start at the suggestion of the Natal amateur botanist Maurice S. Evans, who co-authored the first volume and guaranteed it financially. Later a Natal government grant of £130 per volume allowed Wood to employ Frieda Lauth to draw most of the plates in the early volumes. Several drawings were also executed by Walter Haygarth.[32]

Kew was pleased that Wood had undertaken this task but suggested to him that he insert a reference to each plant stating where it was first botanically described. Wood took much of the information in some of the volumes from *Flora Capensis*. His work was not without its critics and in 1900 the *Journal of Botany* commented, 'We expect more from the man in the field.'[33]

Collectors

FOR DECADES WOOD SET OUT IN a wagon on epic collecting trips in the colony. The best time to collect was the spring but this placed a great strain on Wood partly because of the increasing heat after September and partly because he had to prepare his annual report then for publication early in the new year. It was not until 1901 that the government permitted the issuing of the gardens' grant at the end of June, and from then on the annual reports ran from July to June.[34]

---

29 K.A.: Wood, 30 June 1885.

30 K.A.: Wood, 18 May 1888.

31 K.A.: Wood, 17 June 1898; and Miscellaneous reports: Natal Botanic Gardens, vol 305, Wood to Kew, 27 May 1891.

32 K.A.: Evans, 2 May 1898; and Wood, 28 June 1895, 4 December 1896, 10 June 1898, 17 August 1898, 11 November 1898 and 4 December 1898.

33 *Journal of Botany*, (1900), pp. 192-3.

34 K.A.: Wood, 17 August 1898.

Though he was also in close contact with Dr Rudolph Schlechter of Berlin, Wood's main loyalty remained with Kew. His contributions to the institution and thus to botany prompted Thiselton-Dyer to nominate Wood an associate of the Linnean Society. Wood was elected to the society in 1887. Wood's achievements are reflected in the fact that Ross's *Flora of Natal* lists one genus, the genus Asclepiadaceae, and 60 species named after him.

Wood was sometimes assisted in collecting by James Wylie and, before he became too involved in colonial politics, by Maurice Evans. Other amateur collectors at the time included Captain Albert Allison and his son Martinus; S. Attwood; Mrs Mary Elizabeth Barber and her brother Colonel L.H. Bowker, who was now greatly interested in mushrooms; C.B. Lloyd; Sir Henry William Peek who collected algae; James Ralph; and the previously mentioned Mrs Katharine Saunders and her son Charles. Across the Natal border in East Griqualand William Tyson was for part of the 1880s supplying Kew with interesting specimens from this upland region.

Mrs Saunders was thanked by Thiselton-Dyer in the preface to volume IV of the *Flora Capensis* for the 'interesting plants' she sent to Kew. She fell out with Medley Wood and ignored the existence of the colonial herbarium, sending her specimens for identification to Kew or to Harry Bolus in the Cape. Into the 1890s she continued to request that new specimens should be named in her son Charles' honour. Once he had made his "hit", Mrs Saunders then pressed that her daughter-in-law be commemorated. This was duly done with the epiphyte *Angraecum maudiae*, described by Bolus.

Katharine Saunders also became interested in 'Natal earthworms' and carried on a correspondence with Kew on the subject for some time. In 1889 she gave some folios of her flower paintings to the Natal Society in Pietermaritzburg. By the early 1890s she was less active than she had been, though in 1891 she made the effort of sending seeds of some Natal trees to the British Guiana botanic gardens. The death of her husband in 1892 came as a great blow to her. None the less she still occasionally sent specimens to Kew, the last arriving in 1901, the year she died. The *Kew Bulletin* noted that between 1881 and 1889 she had contributed 426 specimens to the Kew herbarium.

The number of collectors in Natal in the closing decades of the Victorian era had diminished since the heyday of colonial Natal plant collecting in the 1860s. In this later period letters from Natal to Kew, other than those from Wood, tended to ask for garden plants and seeds or, more often, for agricultural plants. Kew was not favourably disposed to these requests and usually asked for plants or seeds in exchange.

Maurice S. Evans (1854-1920)
Evans is probably better known as the Natal politician who supported the local system for Africans. Like many colonial politicians and administrators in the Victorian empire he was also fascinated by plants. With Medley Wood he was author of volume one of *Natal Plants*. In later years he was chairman of the Durban Botanic Society.
(Courtesy: National Botanical Institute, Pretoria)

Katharine Saunders made this watercolour of the water lily *Nymphaea coerulea (N.stellata)*. This attractive plant is still to be found on the pans of Zululand and in isolated ponds throughout Zululand and parts of Natal. It is a plant of tropical and sub-tropical Africa which was described in *Curtis's Botanical Magazine,* (t552), as early as 1812. (Courtesy: Natal Museum, Pietermaritzburg. Saunders' paintings, volume 2, plate 118)

Much of the excitement and dedication attached to plant collecting had passed; its kudos was diminished. In the early 1890s Kew was involved in a row with the great London orchid growers, Sanders and Company, over specimens sent to the firm by Martinus Allison. Sanders was meant to send Kew a proportion of these gratis but due to a misunderstanding Kew was invoiced for the plants. In a letter apologising to Kew for this confusion, Allison's father, Captain Albert Allison, a Natal magistrate, clearly enunciated the attitude of many collectors in the colony in the 1890s:

If it were well known that valuable plants would command a fair remuneration, paid either at your establishment or elsewhere, there would be a great many more specimens sent home - there are no facilities offered to the collector, if he send the fruit of his toil to growers, the latter gets the lion's share - there is practically no reliable market for the collector...You exchange with all colonial curators - the curators depend upon private or amateur collectors, the former have no funds at their disposal to employ and pay the latter...Now and again a German botanist makes his appearance in South Africa, but those I have known do not make "beelines" and have not an unnatural dislike to take the plunge amongst savage life, that is in localities where possibly instead of collecting, they might be collected.[35]

An attempt was made to commercialise collecting throughout southern Africa in the mid-1890s when the Port Elizabeth firm of Laidley and Company established the

35 K.A.: Albert B. Allison, 16 May 1892.

East - South - Central - African Herbarium (E.S.C.A.H.). For ten shillings subscribers received 100 mounted specimens from all over southern Africa. N.E. Brown of the Kew herbarium advised Thiselton-Dyer to subscribe for a time.[36]

Collecting in the extremities of Natal

THE CLOSING YEARS OF THE nineteenth century were not healthy ones for the development of Natal botany. Medley Wood and his *Natal Plants* project survived as a reminder of an age with a greater enthusiasm for botany. Those survivors of the heyday of colonial botanising, Saunders, Sutherland and Wood himself, were now getting old. In South Africa botany was tending to become more the preserve of the professionals; this was especially so in the Cape, which was now reasserting its predominance in botanical matters over Natal. Though the Durban botanic gardens remained unrivalled in Africa, MacOwan and Bolus in the Cape were unquestion- ably the leaders of South African botany; Wood in Natal and Selmar Schönland in the eastern Cape were as provincial caesars to the emperors in Cape Town.

The remote regions in Natal and beyond still offered excitement to botanists of a pioneering disposition. Professor Oliver at Kew considered the Drakensberg 'probably rich in good things.'[37] Maurice S. Evans was particularly interested in the 'marvellously rich flora' of this region. Following a visit to Kew in 1893 when he met Thiselton-Dyer, Evans returned to Natal full of enthusiasm for botany.[38] For several years he made collecting expeditions to the mountains. In 1894 he explored the source and headwaters of the Bushman's River. A year later, as jealous as any prospector for gold or diamonds of the area where he staked his claim, Evans was greatly disturbed when 'a German collector' (perhaps Schlechter) came 'following in my steps in this corner of the colony.' To Thiselton-Dyer he wrote, 'We in Natal are glad to have any help to further knowledge of our flora but would like the credit for any original collecting done by local men.'[39]

In July 1881 a correspondent from the Eastern Cape had written to John Baker at Kew. 'If you could make a trip to Zululand and Pondoland I think you would find some good lilies and parasitical plants worth notice for the country is almost virgin to botanists.' The adventurous collector was particularly attracted by Zululand. Though such early botanists as Plant, McKen and Gerrard had ventured there, they tended to follow roughly the same route: the traders' and hunters' wagon road up to and across the Tugela, then pushing north through the bush country between the highlands and the coastal swamps. Very few ventured north of St Lucia into Maputaland. The major problem was malaria and most collectors who entered Zululand suffered from the disease. John Sanderson's brother Septimus had been fortunate in the late 1860s when he escaped death from malaria in Zululand, though he was saddened that his sickness caused him to abandon a 'very remarkable... greenish grey shiny and smooth' *Adenium* in the north of the region.[40]

Miss M.C. Owen had spent under six months in Zululand in the 1830s and had done some collecting, her specimens finding their way to William Harvey in Dublin. A lady collector in Zululand was a very rare phenomenon. Forty-four years after Miss Owen and her missionary brother had fled the upheavals of Dingaan's Zululand, a Mrs McKenzie collected in the south of the region. She journeyed from Richards Bay up the Mhlatuze river valley and on to Isandhlwana.

36 K.A.: E.S.C.A. Herbarium, f856, n.d., South African letters, vol 190.
37 K.A.: Kew memorandom, f1666, 6 January 1884.
38 K.A.: Evans, 27 July 1894.
39 K.A.: Evans, 15 March 1895.
40 K.A.: Sanderson, 22 January 1868.

According to a letter she sent to Kew, the geraniums, lilies and grasses were beautiful in this area.[41]

It was not until after the Anglo-Zulu war of 1879 and the British annexation of Zululand in 1887 that any regular botanising expeditions were made in this region. The fact that there were now magistrates in Zululand greatly helped the process of botanical exploration as did the appointment of forestry officers in the region in the 1890s. Katharine Saunders' eldest son, Charles, was the senior colonial official in Zululand both before and after the colony was incorporated into Natal in 1897. He served at Ubombo, in the north near the Mkuzi river and Lebombo mountains, and at Eshowe, the capital of Zululand, in the south. Under pressure from his mother he sent many plants to Kew and to Bolus in the Cape, including a *Pachypodium* that was named in his honour.[42]

Medley Wood began his regular expeditions to Zululand only in the late 1880s. In 1884 he had turned down an invitation to go to Eshowe, telling Kew: 'Native affairs are in such an unsettled state, that I do not think it advisable to accept at present as any extensive botanising would I fear have to be done under protection of an armed escort.'[43] From 1887 he began to visit Zululand, on one occasion receiving help in collecting from the celebrated white chief John Dunn. Wood's greatest triumph in collecting in Zululand came in 1895 when he discovered a new cycad, *Encephalartos woodii*, in the Ngoye forest. He removed stems of this to the Durban botanic gardens where they can still be seen. No further specimen was ever discovered so eventually the whole plant, including its offshoots, was removed from Ngoye by James Wylie, who by the late 1890s was increasingly helping Wood by going out and collecting for him.[44] Wood also received specimens from Zululand from H. Jenkinson, H. Swanfield and the forest officer, G.H. Davies.

The years of decline

ONE OF THE PRINCIPAL REASONS why these remote regions began to be explored more at the close of the Victorian period was that agriculture was expanding rapidly by the turn of the century. In addition Natal suffered from a series of unfortunate episodes and natural disasters in the 1890s. An attempt to establish a forestry department between 1891 and 1893 foundered due to lack of funds and the personality of the conservator of forestry, Friedrich Schöpflin. Though Kew tried to revive the idea by publishing an article on Natal's forests in 1895, it was not until 1902 when T.R. Sim was brought in from the Eastern Cape that a workable forestry policy was formulated.

Compared with what was to come, the failure of the forestry policy was a minor setback for the botanically minded. Drought returned and by 1899 the annual rainfall in Durban was once more down to 28 inches (71 centimetres) from the average of 40 inches (102 centimetres). This drought exacerbated a problem with locusts and a more serious one of rinderpest. Natal farmers on the whole acted swiftly in erecting fences to prevent the disease being spread by straying infected cattle, but Natal was still badly hit. Here, as the governor pointed out to Thiselton-Dyer, there was the problem of large numbers of cattle, representing the wealth of the Nguni.[45] There was localised famine. With the help of Wood, Hely-Hutchinson despatched to Kew details of plants used as food in these areas. He observed to Thiselton-Dyer, 'The subject is of interest as showing how a native race can keep itself going in these parts, even under the most ad-

---

41 K.A.: McKenzie, 6 November 1882.

42 K.A.: C. and K. Saunders, 5 May 1891 and June 1892.

43 K.A.: Wood, 8 April 1884.

44 K.A.: Wood, 10 May 1895; and *Kew Bulletin*, (1914), pp.250-1.

45 K.A.: Hely-Hutchinson, 8 September 1896 and 24 September 1896.

The entrance of the Jubilee Conservatory, Durban. In 1897 a conservatory was erected in Durban to commemorate the diamond jubilee of Queen Victoria. It cost some £3000 and was situated in the Durban botanic gardens. It was assembled from a 'kit' sent out by a Glasgow firm of James Boyd. This handsome structure was opened amidst great pomp in December 1898 by the governor Sir Walter Hely-Hutchinson.

verse circumstances'. Thiselton-Dyer promised a note on the subject in the *Kew Bulletin*.

Zululand also had the serious problem of tsetse fly. An army officer named Bruce was seconded to Ubombo in Maputaland to investigate the fly. Hely-Hutchinson sent his report to Thiselton-Dyer who forwarded it to the Royal Society. So impressed was the society that a special committee was appointed to investigate the problem. The governor was delighted and wrote to Kew, 'I feel that I owe this to your kind interest, and am correspondingly grateful'.[46]

46 K.A.: Hely-Hutchinson, 21 January 1898.

But an even more serious problem than drought, locusts, rinderpest or tsetse fly lay ahead. The worsening political crisis between the Transvaal Republic and the British threatened the security of Natal. On 4 October 1899 Hely-Hutchinson told Thiselton-Dyer, 'According to reports from Pretoria, we were to have been attacked yesterday or today, but personally I do not think it likely'. Seven days later the Boer forces crossed into Natal and the second Anglo-Boer war had commenced.

The next three months were very dark ones for the beleaguered colony. By early November Wood reported to Kew, 'An hour ago the news of the evacuation of Colenso came to hand, half of Natal is now in the hands of the Boers...I hope that we shall not see them in Durban, but I do not feel at all sure of it at present.'[47] As civilians had to have a pass to leave Durban and none was allowed beyond Pietermaritzburg into the military zone, plant collecting became impossible.

The war years, from 1899 to 1902, were unsatisfactory ones for Natal botany. One major success for the Durban botanic gardens was the flowering in 1901 for the first time in Natal of the giant *Victoria* water lily, grown in a specially constructed lily pond. John Sanderson had hoped for this as early as 1852[48] and Mark McKen had tried repeatedly throughout the 1860s to germinate seeds of the plant sent by Kew, all with no success.[49]

In 1902, the year the Victorian era drew to a close, a purpose-built herbarium was opened in the Durban botanic gardens but it was constructed too late to take advantage of the by now rapidly vanishing colonial plant craze. It was several decades later before popular interest in botany resurged, despite the efforts of such intrepid newcomers to Natal as August Rudatis and T.R. Sim.

In 1903 Wood was appointed director of the Durban botanic gardens, thus allowing Wylie to become curator. The fortunes of the gardens were not good: crippled financially after the Bambatha insurrection of 1906, they were eventually transferred to the municipality in 1913. Wood remained as director of what was now called the Natal Herbarium, under the Union government's division of botany and plant pathology. He died in August 1915 aged 87. Though he had not retired from his botanical enterprises, his contact with Kew had greatly diminished, as had all contacts between Kew and Natal. Wood's relationship with Thiselton-Dyer had not always been good[50] and Thiselton-Dyer, who in 1892 had reprimanded Wood for grumbling in his letter to Kew for the effort involved in collecting, admitted privately to Brown that he had not even read a note concerning Wood in the *Gardeners' Chronicle*. Thiselton-Dyer was also less willing to help with enquiries from institutions which were not botanical stations. When A. Pearson, the director of the Natal department of agriculture, wrote to him about a problem in 1902, Thiselton-Dyer replied, 'Agricultural work in the ordinary sense does not fall within the scope of Kew': a very different response from that given to similar enquiries received from Natal a generation earlier.[51]

IN SOME RESPECTS KEW itself was entering into a period of decline at the turn of the century, though the staff employed was nearly 200 and the number of visitors approached 1 000 000 annually. The *Kew Bulletin* was ridiculed as being nothing now but appendices. Thiselton-Dyer was notoriously quarrelsome and, following a row with the gardeners in 1905, he resigned.[52] Two years previously Kew Gardens had been transferred to

---

47 K.A.: Wood, 3 November 1899.

48 K.A.: Sanderson. 9 November 1852.

49 K.A.: McKen 1 January 1866, 10 August 1866, 11 January 1867, 10 May 1867, 10 November 1867 and 19 September 1870.

50 See K.A.: Wood, 28 April 1892, 18 May 1893, 24 May 1893, 27 September 1893, 17 August 1898, 20 January 1899, 17 March 1900; and Kew memorandom, 12 June 1893.

51 K.A.: Natal miscellaneous, 1862-1909, vol 305, Pearson to Thiselton-Dyer, 17 May 1902, and Thiselton-Dyer to Pearson, 14 June 1902.

52 F. Nigel Hepper, *Royal Botanic Gardens, Kew : Gardens for science and pleasure*, (H.S.O., 1982), p.17; and Turrill, *Kew: Past and present*, p.35.

the board of agriculture and fisheries.

That the new director of Kew, Lieutenant Colonel David Prain had come from a colonial institution, the Royal Botanic Gardens, Calcutta, was a reflection of the success Kew had had over the previous generation in promoting colonial botanic gardens and stations. Prain had heard of the work done over the years in Natal and had exchanged specimens with Wood. In 1913, the year Wood was awarded an honorary doctorate by the university college in Cape Town, the National Botanic Gardens at Kirstenbosch was established. Prain tried to prevent the downgrading of the Durban botanic gardens, but without success. In 1913 the *Kew Bulletin*, in a short but sincere appeal, marked the end of a richly productive relationship between what had been a small colony and the world's finest botanical institution:

It must not be forgotten before passing on to the consideration of the National Botanic Gardens at Kirstenbosch that in Natal South Africa has possessed a botanic gardens for over fifty years where the true functions of such an institution have been ably maintained in spite of many difficulties. It is a matter of regret that the area of this garden is so small, but small though it be its maintenance is as important now as ever it was, and its activities must not be suffered to be curtailed or its functions abrogated owing to any change in its administration or to the establishment of the new National Gardens.[53]

A pathway in the Pietermaritzburg botanic gardens. This garden was noteworthy for the large number of trees grown there. The distribution of thousands of young saplings prompted praise from Sir Joseph Hooker.
(Courtesy: Killie Campbell Africana Library)

---

53 *Kew Bulletin*, (1913), p.309.

Chapter 6

Later plant exchange

THE COLLECTING OF PLANTS IN Natal was considerably facilitated by the expansion of the railway into the interior. In 1885 it had stretched to Ladysmith and by 1891 Charlestown on the colony's northern border. Five years later the connection between Johannesburg and Durban was complete. The progress of the railway along the north coast was less speedy and for many years the railhead had reached no further north than Verulam.

Vast tracts of the colony still lay far from any railway station and the traditional practice of taking a wagon and slowly trundling over the veld, sometimes for weeks on end, had to be resorted to for collecting in Zululand, Alfred County and the Drakensberg. Usually such parties consisted of the collector and a few African helpers. An exception was Medley Wood, who in the heyday of his collecting in the late 1880s often took his wife and a large party of helpers along, the entourage taking from a month to six weeks to complete an expedition.[1]

With these traditional expeditions there were all the problems encountered by the early collectors. The flooded Tugela River trapped Wood in Zululand in 1886 and forced him to make a long detour upstream to cross near the source. A year later he failed to reach the Nkandhla-Qudeni forest region because of heavy rains. In 1889 he had a very different problem when he spent too long collecting in the bright sunlight and suffered badly afterwards with 'inflammation of the eyes'.[2]

Even acquiring a wagon for the expedition could be a problem for the collector. Both John Sanderson and Medley Wood usually hired a vehicle for their trips. But when there was an increase in the trade across the Drakensberg mountains or a rush to get to the diamond mines or goldfields in the interior, wagon hire or even purchase became very difficult in Natal. Returning wagons from the diggings might be persuaded to carry plants to Pietermaritzburg or Durban but even an ordinary cart charged four to five shillings to carry plants into Durban from not very far away.[3]

On the next part of their journey overseas to Britain, plants could be carried much more swiftly than before. By the mid-1870s the mailship could be expected to take 26 days from Cape Town to England. Another week and a half had to be allowed for the leg from Durban so the whole journey from Durban to England took five to six weeks. By the 1890s this was reduced to three weeks, which was of considerable help to Natal botanists who could now have a specimen identified by the Kew herbarium and have a report on it before a change of season.

---

1 K.A.: Wood, n.d. (c.1887) and 5 September 1895.

2 See K.A.: Wood, 1 March 1886, 17 October 1887 and 9 January 1889.

3 See K.A.: Wood, 6 October 1882, 27 February 1889 and 18 March 1890; and Sanderson, 3 February 1869.

The general question of season was raised by several collectors from the mid-1860s. Experience had shown that many plants did not survive if sent into a European winter. But conversely the best time to despatch living plants was in their dry and dormant period during the African winter months. Wood considered September to be too late to send plants to Europe.[4] He had tremendous trouble sending live specimens of *Gerrardanthus* to Kew for this reason and despite pressure from Kew he wrote to Joseph Hooker in 1880:

Our spring has now fairly set in, and in my opinion it will be useless to attempt sending it before next winter. I saw some of the plants yesterday and the growth has fairly commenced.[5]

WHEN SPECIMENS WERE EXCHANGED between Natal and Kew they were still often despatched in wardian cases. Glass breakage remained a serious problem. When one Natal recipient, named Blackburn, found a quarter of the plants that Kew had sent him had died of exposure because glass panes were broken in the case, he wrote to Kew and asked that in future they cover the case with wire netting which was the normal practice.[6]

Mark McKen once complained that some azaleas he had received looked 'seedy' because they were in a wardian case 'which had just been painted, as they appear to have absorbed a quantity of oil'.[7] More frequent complaints were still that the plants had been over or under watered, that they had been improperly packed or that they had been attacked by insects.[8] Even in herbaria there was the hazard of attack by the drug-store beetle, *Stegobium paneceum*. Wood firmly believed that seeds were safe from attack only if they were in a hermetically sealed packet.[9]

As the century progressed increasing use was made of dry boxes to transport smaller specimens. A dry box was smaller than the usual ocean-going wardian cases. It was used to transport live bulbs and roots and such species as orchids, cacti and other succulents which were often packed in moss.[10] In 1867 McKen commented:

I find that in sending away succulents in dry boxes it is well to ventilate them by boring a few small holes in the top and bottom of the case. This prevents fermentation by over heating during the passage.[11]

John Sanderson used a larger box to send *Euphorbia* to Kew. He lined the container with zinc or tin to prevent attack by rats.[12] Bottles were also sometimes used to send Kew flowers and fruits of species preserved in spirit.[13]

The use of the official government despatch bag by Natal collectors to send specimens free to Kew came under attack in 1875. In that year a parcel of dried plants Sanderson was sending to Hooker was 'inadvertently opened' by the Natal colonial secretary. 'This insolent jackanapes' not only forbade future use of the bag for this purpose but tactlessly ventured to instruct Sanderson how to parcel such specimens. A series of heated letters was sent to the colonial secretary and the governor by Sanderson who, having commented on several spelling mistakes in the offending letter, told the colonial secretary not to try to 'teach his grandmother to suck eggs'. Kew was about to enter the fray when Sanderson was sent a letter of apology by the governor. In future, however, the civil service was reluctant to accept large parcels for forwarding to Kew by this means, Medley Wood in 1880 and Colonel Bowker in 1884 both received warnings in this regard.[14] However in the 1890s the government allowed specimens sent to Medley Wood for work on his *Natal Plants* to be carried free of charge within the colony.[15]

4 See K.A.: McKen, 10 November 1866, 14 December 1868 and 12 January 1869; and Wood, 24 September 1883.

5 K.A.: Wood, 14 October 1880 and 26 July 1880.

6 K.A.: Blackburn, 6 April 1869; and see Keit, 19 December 1872.

7 K.A.: McKen, 10 February 1867.

8 K.A.: Topham, 9 August 1870 and 11 June 1873; and Wood, 23 July 1879, 5 April 1880 and 11 November 1882.

9 K.A.: Wood, 8 February 1883.

10 For example, K.A.: Keit, 27 November 1877; and Sanderson, 7 March 1868.

11 K.A.: McKen, 11 January 1867. See also McKen, 6 July 1868.

12 K.A.: Sanderson, 3 February 1869.

13 For example, K.A.: Wood, 5 April 1880.

14 K.A.: Sanderson, 3 November 1875, 15 November 1875, 16 November 1875 and 10 December 1875; and Wood, 22 December 1879 and 5 April 1880.

15 J. Medley Wood, *Colonial Herbarium, Report for the year 1898*, (Durban, 1899), p.4

VERY OFTEN SHIPS' CAPTAINS went to considerable trouble and personal inconvenience to ensure that a case was well looked after on board ship. This did not apply only to captains of the Union Packet Company, which carried wardian cases free of charge, but also to independent 'liberal shippers' whom McKen sometimes rewarded with a small case of bulbs for their troubles.[16] Though it is not known in this instance if their problem arose in 1868 when a steamer containing a Kew wardian case was unable to communicate with the shore because of bad weather.[17]

Many cases were still accompanied by travellers keen to have an excuse to call on the director of Kew. By the mid-1860s carriage for a wardian case could vary from £3 to £6. Unaccompanied on the voyage, a case could easily be forgotten about. In 1867 McKen reported to Joseph Hooker: 'The case Mr Smith [curator at Kew] last sent with orchids etc was by some mistake taken on to Ceylon, the steamers in their hurry often do this sort of thing'.[18]

Increasingly the transport of plants became more formal with cases having bills of lading and parcel tickets being issued. Confusion over this procedure often arose when the case's papers were lost and the case left unclaimed at the docks. This is what had happened to the case of Uba cane mentioned in a previous chapter, which Wood had stumbled across at the Durban docks. In 1890 by a strange coincidence it was a Kew man who saved two wardian cases from a similar fate in Durban. Wood explained to Thiselton-Dyer:

Unfortunately the B/L [bill of lading] did not come to hand, and the cases were on the [Durban] wharf a week before I knew of their arrival. They might have remained several days longer but for the kindness of a man from Kew on his way to Quilimane, who saw them on the wharf and took the trouble of walking up here to let me know.[19]

Despatches from Natal to Kew

IN THE PERIOD FROM 1866 to 1902 over 90 wardian cases or dry boxes were despatched from Natal to Kew. A large number of parcels was also sent. During these 36 years the greatest volume occurred in the late 1860s when 26 consignments of cases, boxes and ordinary parcels were despatched to Kew. *Johnson's Gardeners' Dictionary*, 1894 edition, lists 75 Natal plants newly introduced to Europe: 28 percent are listed as being introduced before 1866, 57 percent in the period from 1866 to 1882, and 15 percent in the period from 1883 to 1894. While the plants listed are only a small proportion of the total number of Natal novelties sent to Europe, none-the-less the statistics give a fairly accurate picture of the fortunes of both plant collecting in the colony and of plant exchange with Kew. The heyday is clearly in the late 1860s and 1870s, despite the problems facing the Durban botanic gardens at this time. While Keit complained in 1873 that his gardens had nothing Kew 'might care to have',[20] the following year he was able to send Hooker 20 cycads and some tree ferns.[21]

In the 10 years from 1871 to 1880 some 44 percent of wardian cases or dry boxes received by Kew from Africa came from Natal. Figures of the number of packets of seed despatched from Natal to Kew do not exist, but a steady stream was sent until the end of the century. In addition to Natal plants, Kew received specimens which had been forwarded from Durban after being collected outside the colony in regions such as East Griqualand, the Transkei, the Transvaal, Orange Free State, Swaziland, Portuguese East Africa and as far afield as Namaqualand. In 1883, for example, arguably the most beautiful gladiolus, *G. oppositi-*

---

16 K.A.: McKen, 15 December 1865, 3 June 1866, 13 August 1866, 10 November 1866, 10 November 1867 and 7 December 1867, 16 July 1892.

17 K.A.: Sanderson, 6 January 1868.

18 K.A.: McKen, 10 November 1867.

19 K.A.: Wood, 22 October 1890.

20 K.A.: Keit, 23 September 1893.

21 K.A.: Keit, 6 October 1874.

florus (*G. salmoneus*) was discovered near Kokstad in East Griqualand and sent to Kew via Natal. Similarly the *Pterodiscus speciosus* from the Transvaal was sent to Kew by Medley Wood in 1886.[22]

In the 1860s ferns and tree ferns were much in demand in Europe. The latter had been one of the major features of the 1866 London international exhibition. *Alsophila capensis, Cheilanthes capensis, C. hirta, Lomaria cycadioides, L. zamaefolia* and *L. zamioides* were all listed by one contemporary work as being South African ferns which were greatly prized by British collectors.[23] Similarly the two known Natal palms, the wild date palm, *Phoenix reclinata*, and the lala palm, *Hyphaene natalensis*, were sought after overseas. Other plants in demand were any species of succulents, orchids, *Euphorbia, Gladiolus* or *Hibiscus*.

WHEN THISELTON-DYER WAS director of Kew he occasionally sent out lists of 'Desiderata' to imperial botanic stations and gardens. As cycads were his 'special pet', Natal was an important region in his eyes and many specimens from the colony found their way to the great palm house. In the closing decades of the nineteenth century, before the *Encephalartos woodii* was discovered, Kew listed those cycads which had been received from Natal as: '*E.barterisis, E.brachyphyllus, E.caffer, E.ghellinckii* and *E.villosus*'.

Thiselton-Dyer was also interested in aloes and the plant lists and Natal-Kew correspondence make reference to such species as *A. bainesii, A.cooperi, A.kraussii*, and *A.arborescens*. A consideration for collectors in sending plants to Europe was whether Natal species would grow there. The spectacular Natal flame bush *Alberta magna, Littonia modesta* and *Buttonia natalensis* were a great challenge to Kew in its attempt to cultivate them in England. *B. natalensis* in particular proved very difficult to grow.[24] On the other hand, the purple-flowering pineapple lily, *Eucomis bicolor*, was by

ALBERTA MAGNA (Natal flame bush) Lieutenant Governor Scott found this small tree growing out of a precipice at Karkloof in 1863. Dr Sutherland was unable to identify it so he sent samples from the species to Kew in the official government despatch bag. The plant did not flower at Kew until 1895 and then it was from seeds sent from Natal in 1889. This handsome tree is a member of the Rubiaceae family. (*Curtis's Botanical Magazine*, T7454, [1896])

Bottom left:
ALOE BAINESII The famous tree aloe which was described by Thiselton-Dyer and named after Thomas Baines, one of South Africa's most fascinating explorers. (*Curtis's Botanical Magazine*, T6848 [1885])

22 K.A.: Wood, 28 July 1886.
23 B.S. Williams, *Select ferns and lycopods*, (London, 1873).
24 K.A.: Wood, 18 March 1892 and 28 June 1892; and J. Medley Wood, 'Revised list of the flora of Natal', *Transactions of the South African Philosophical Society*, XVIII.2, (1908), p.201.

1884 being grown along the outside walls of Kew's great palm house.

Also in 1884 attempts were made by Natal collectors to introduce to Britain two possible new garden flowers. Colonel Bowker sent Kew seeds of what was thought to be a species of *Scilla* : 'Black seeds like rifle powder - is a waterbulb and very pretty, and I think it would grow in the open in the south of England'.[25] Two months previously Dr Sutherland had sent Kew seeds of what he described as a red daisy-like *Lobelia* which was 'carpeting the hills in Natal in fields where the soil had been exhausted'. He noted of it, 'I think it may thrive in England as its near relative the blue one does'.[26] One of the most popular Natal plants in late Victorian Britain was the dwarf perennial succulent, *Huernia hystrix* or the porcupine *Huernia*. This had first been sent to Kew by McKen in 1869.

A new development in the 1880s was the despatching to Kew for identification of exotic invaders in the colony. An American species of *Phytolacca* was particularly troublesome.[27]

THE FLOW OF LIVE SPECIMENS to Kew in the closing decades of the nineteenth century was complemented by a steady stream of material sent to the Kew herbarium. Between 1878 and 1899, for example, Wood sent the herbarium over 4 000 specimens, a volume similar to that despatched to Kew by the great Cape botanists, Bolus and MacOwan.

Increasingly the staff of the herbarium were not content with a single sample and a few rough notes on a specimen. Not only did they insist on proper numbering of specimens but they frequently requested a collector to send male and female specimens of the same species. They also requested fruits and seedpods and detailed information on the colour and size of the species growing in the wild. There were numerous problems encountered in identification. A good example of this was one Natal specimen which caused Barker much trouble in the early 1880s. To Medley Wood he wrote of it:

Hibiscus physaloides...I previously wrongly named this H. vitifolius, but it seems to me that the plant Harvey calls H. natalensis is the true H. vitifolius, but it wants further examination than I have been able to give it.

Results of Kew's identification were sometimes surprising. Concerning a *Hydrocotyle* var. *monticola* from Natal, Barker noted:

A very interesting discovery as it has only been found in Fernando Po [an island at the foot of South America] at 8 500 feet. And although the two localities are so widely separated the two plants are absolutely identical.[28]

Two categories of exchange between Natal and Kew increased significantly in this later Victorian period. One was plant photography: the advantage of this to such photographers in the colony as Kisch was that as long as there was no strong wind the subject of the exposure remained still. As early as November 1867 McKen was writing to Joseph Hooker:

I beg you will accept of the enclosed photograph of a giant Encephalartos for the Kew museum and I shall be glad of one of yourself in return - one of your late curator [John Smith] would also be very acceptable.[29]

From the late 1860s Kew received botanical photographs of Natal species from such collectors as Sanderson and Wood.[30]

The second category of material despatched was one way, from Natal to Kew, and related to African life in the colony. Though John Sanderson might write, 'I fear many of our people are nigger haters at heart', none the

---

25 K.A.: Bowker, 19 July 1884.

26 K.A.: Sanderson, 6 May 1884.

27 For example, see K.A.: Sanderson, 6 September 1880 and 11 November 1882.

28 Natal Herbarium: Kew list, October 1880 to April 1881 and August 1881 to February 1882.

29 K.A.: McKen, 10 November 1867.

30 For references to botanical photography, see K.A.: McKen, 10 March 1869; Sanderson, 3 February 1869, 22 August 1869, 2 March 1874 and 31 July 1875; Wood, 29 December 1882, 18 May 1888, 24 August 1894, 15 March 1895 and 21 July 1899; and *Kew Report* (1877).

less many botanists were interested in African customs and life.[31] They were encouraged in this by Kew, which in 1881 opened a new wing to its museum number I.[32] J.R. Jackson, the museum curator, was keen on acquiring information and artefacts relating to 'tribal life'. If there was a 'botanical' connection so much the better. Thus an African necklace made from a 'species of Natal *Protea*' which Katharine Saunders sent to Kew was gratefully received.[33] In 1881 Thiselton-Dyer wrote to Bowker, 'We much want for the museum native spears with handles of the South African bamboo'.[34]

Kew was, however, keen to get any implements relating to Nguni life for its museum.[35] In this it was greatly assisted by Bowker who sent stone implements from the 'Umbolvu river near St Lucia',[36] as well as by John Sanderson and P.C. Sutherland. Sutherland sent African mats called 'Goloto' woven from an indigenous grass.[37] Sanderson sent a large collection of material, including blankets made from the bark of the 'Bashlapo', stone weapons, snuff boxes, pipe heads, a crucible and stone blacksmith's hammer, clay figures of animals and people, and a bottle made from untanned skin.[38] Of the last of these he wrote, 'I had no idea the natives were up to such art'.[39] Most collectors were interested in such things and few were as dubious of the value of looking for such artefacts as A.S. White who noted of Bowker's discovery of stone implements, 'I am rather sceptical of these'.[40]

Some collectors had a wider interest than simply gathering artefacts. Sir Theophilus Shepstone and Miss Colenso both worked on food crops grown by the Nguni and sent lists of their local names to Kew. As has already been seen, White told Kew of the 'mundi' plant much eaten by the amaPondo. Wood studied a species of Labiatae, *(Coleus)*, known as the 'kaffir potato', which he believed had been introduced into Natal from Zululand.[41]

Both Medley Wood and Kew were sceptical about the efficacy of African medicines derived from plants, despite claims that *Rauvolfia caffra* could cure malaria and *Turraea floribunda (T. heterophylla)* dysentery. On one occasion sap of *Sarcostemma viminale*, alleged to relieve pain if the poisonous latex of the tambotie tree *(Spirostachys africana)* got in the eye, was sent by Thiselton-Dyer to a London medical specialist, E. Nettleship. He reported back:

The results so far as we went were quite negative. The sample sent would seem to be greasy, at any rate it does not mix with the tears and when put inside the eyelid but behaves like any oily liquid would do.[42]

Wood himself had no time for Maurice Evans' suggestion that they should write a book on 'Native plant medical remedies'.[43]

Despatches from Kew to Natal

THOUGH IN THE LATE 1860s and early 70s the emphasis was on plant exchanges, as the nineteenth century moved to a close the despatches of material from Kew to Natal shifted from plants to mainly seed. Between 1866 and 1875 31 wardian cases were sent to Natal by Kew. They contained in excess of 800 live specimens. Between 1871 and 1875 62 percent of wardian cases despatched to Africa were sent to Natal. Precise statistics for the number of packets of seed do not exist for the whole period but it is known that in 1872 Kew sent the colony 432 packets, being 73 percent of the amount sent to Africa that year. In 1874 this proportion had dropped to 15 percent, thus illustrating the fluctuation in the exchange. Taking the average for the years when statistics do exist, about

31 K.A.: Sanderson, 2 March 1874.

32 Turrill, Kew: Past and present, p.31.

33 K.A.: K. Saunders, c. 10 August 1881.

34 K.A.: Bowker, 16 November 1881. See also W.H. Flower, 12 June 1880.

35 See *Gardeners' Chronicle*, (1890), p.381; K.A.: Bowker, c. 1881 and 30 May 1882; Sanderson, 7 March 1868; Topham, 18 July 1869; and Wood, 4 December 1882 and 8 January 1883.

36 K.A.: Bowker, 4 May 1880.

37 K.A.: Sutherland, 23 April 1869.

38 K.A.: Sanderson, 8 March 1866, 8 January 1869, 22 August 1869 and 11 October 1869.

39 K.A.: Sanderson, 15 February 1870.

40 K.A.: White, 4 March 1868.

41 K.A.: Natal miscellaneous, 1862-1909, ff304-8; Keit, 17 January 1873; and Wood, 28 July 1886.

42 K.A.: Nettleship, January 1892.

43 K.A.: Wood, 26 August 1891.

3 250 packets of seed were sent to Natal in the period from 1866 to 1875.

Whereas the flow of material from Natal to Kew from 1876 to 1881 continued relatively normally despite the problems in the colony, the late 1870s see a cessation of plants and seeds being despatched by Kew to Natal. This was due to the decline in fortunes of the Durban botanic gardens which received most of the wardian cases arriving in the colony. It was not until 1882 that the flow from Kew recommenced.

From the early 1880s the Durban botanic gardens was added to the general Kew seed distribution list of imperial botanic stations. This was obviously of great benefit both to the gardens and to the colony. The following are some examples of species distributed to Natal and the empire by Kew, as listed in the exchange volumes:

1882 *Parmestera cirifera*
1883 *Musa ensete* [banana species]; *Pyrethrum cinerariaefolium* [chrysanthemum species]
1886 Havana tobacco; *Acacia gummifera*
1887 *Eragostis abyssinica*
1889 *Copernicia cerifera*
1897 *Melhania* sp.
1898 *Oxalis crenata* [wood sorrel species]; Trinidad white yam; *Villubrunea integrifolia*
1899 *Hesperaloe engolmannis*
1901 *Arundessaria falconeri* [tomato species]

The stations receiving such seed from Kew included Accra, Adelaide, Antigua, Bangalore, Baroda, Berlin, Birmingham, Brisbane, Calcutta, Cambridge, Cape Town, Ceylon, Demerara, Dominica, Edinburgh, Fiji, Gambia, Glasgow, Glasnevin, Grenada, Hanover, Hobart, Hong Kong, Jamaica, Java, Lagos, Leiden, Lucknow, Madras, Mauritius, Melbourne, Montserrat, Natal, Ootacamund, Oxford, Paris, Poona, Port Darwin, Saharanpur, St Kitts, St Lucia, St Vincent, Sierra Leone, Simla, Singapore, Sydney, Trinidad and Trinity College, Dublin.[44]

Such material was often supplied to Kew by colonial institutions or by the British foreign office. Occasionally seed received from such sources was despatched specifically to Natal. Thus through the agency of Kew Natal received: *Boswellia serrata* and *Phoenix paludosa* (swamp date palm) from Calcutta (1883 and 1887 respectively); *Haloxylon ammodendron* from St Petersburg in Russia (1888); *Acacia pycnantha* (golden wattle) from Sydney (1890); *Melhania erythroxylon* from St Helena (1892); *Coffea stenophylla* from Sierra Leone (1894); *Cola acuminata* from Jamaica (1894); and *Licuala grandis* from British Guiana (1894).

Kew records make no reference to the *Jacaranda mimosifolia* which was to become a notable feature of Durban and Pietermaritzburg streets in the twentieth century. Tradition has it that this species was introduced by James Wylie who came to Durban botanic gardens from Kew in 1882.[45] Medley Wood in his 1897 guide to the gardens cites one specimen as having been planted in 1885.[46] The jacaranda had been introduced to Europe from Brazil in 1818 so it was certainly possible that one or more of the many packets of seeds unnamed in the exchange books which were despatched to Natal in the 1880s contained this species.

Another tree that was to become popular in the colony by the turn of the century definitely owed to Kew its introduction into Natal. From the mid-1860s Mark McKen began to request Kew to send him species of *Araucaria*. As well as the monkey puzzle *(A. araucana)*, these included bunya

---

44 K.A.: Exchange book, 1896-1923, 1896.

45 *Daily News (Natal)*, 2 July 1947.

46 J. Medley Wood, *Guide to the trees and shrubs in the Natal Botanic Gardens*, (Durban, 1897), p.34.

bunya *A. bidwillii* and Moreton Bay Pine *(A. cunninghamii)*. Equally successful at adapting were the specimens of deodar cedar *(Cedrus deodara)* and various species of cypress sent by Kew. Less promising were the giant redwoods *(Sequoia gigantea)* and the Norfolk Island pine *(Araucaria heterophylla)* brought from Kew in the late 1860s. Robert Topham and A.S. White were especially keen on growing the latter but found it extremely difficult to germinate seed. White was of the opinion that this was because Kew did not send the seed in the cone and, as has been seen, presumed to lecture Joseph Hooker on the subject.

Flowering garden plants which were sent by Kew in this period and grown successfully in Natal include varieties of *Azaleas*, cactus, *Fuchsia*, orchid, *Pelargonium*, and *Rhododendron*. Kew is also recorded as having sent such attractive species as *Bignonia picta, Exbucklandia populnea, Gardenia fortunei, G. intermis, Hibiscus pedunculatus, H. schizopetalus, Maquaira grandiflora, Passiflora maliformis, P. racemosa, Pavonia multiflora, Picrasma excelsa, Pseudophoenix sargentii, Serenoa serrulata* and *Villabrunea integrifolia*.

The despatch of 'economics' to Natal continued, though Kew was now reluctant to accede to all requests as certain types of plantation crop seed could now be obtained from botanic institutions and nurserymen both in Natal and the Cape. None the less special requests were granted, such as that from W.R. Hindson and Company of the Natal Clifton tea estate who wrote, 'We are growing tea extensively, but should much like the strain to be improved. We tried to get some over from India last year, but it got completely killed in transit'.[47]

Literature

FROM THE MID-1860S Kew despatched to other botanical institutions an increasing volume of printed material on botanical subjects. In 1872 Sir Joseph Hooker commented to John Sanderson that he was sending books to 'a man who has made a name'.[48] Kew obviously held Natal botanists in high esteem for a large quantity of printed material was sent from the institution to the colony. In addition Kew provided lists of suitable botanical works for those who asked for guidance in this matter. A good example of this was the Rev. W.A. Newnham, headmaster of Hilton College, who in 1874 requested information on up-to-date works to assist him in his teaching of natural science in the school.[49] Kew also selected and bought botanical books for regular correspondents who sent money for this purpose.[50]

Most of the literature which Kew despatched to Natalians was sent as a gift. The following publications are mentioned in Natal-Kew correspondence as having been presented to various colonists by Kew: articles by Thiselton-Dyer on aloes; *Cultural Industries;* the later parts of the *Flora Capensis* and parts of the *Flora of Tropical Africa* and various parts of the flora of British India; the *Gardeners' Chronicle; Grevillea; Icones Plantarum; Index Floral Sinensis;* Joseph Hooker's address to the British authorities in 1896, his paper on insular flora and his address to the Royal Academy; the 'memoirs' of Sir William Hooker; the *Kew Bulletin,* annual reports on the gardens and the official guide to the Kew museum; *Paxton's Botanical Dictionary; Report of the [British] Board of Agriculture, Report of the experimental farm at Ottawa; Report of the gardens at Hong Kong; Report on the Madras Exhibition* (c.1884); and *Synopsis Filicum.*

In 1877 William Keit advised Kew that the best means of sending printed material to the colony was to post it care of P. Davis and Sons, who printed among other things the *Natal Witness.*

47 K.A.: W.R. Hindson, 19 June 1890.
48 K.A.: Sanderson, 11 December 1872.
49 K.A.: Newnham, 9 April 1874.
50 For example, K.A.: Wood, 6 April 1886 and 9 November 1886.

According to Keit, they had parcels and cases arriving in Natal by every mailship.[51]

The nurserymen

WILLIAM KEIT, LIKE T.R. SIM a generation later, and a number of other curators of Cape botanic gardens, was to move from running a botanic gardens to owning and managing a commercial nursery. By the 1880s a thriving nursery industry was developing in Natal. By the 1890s, under the pugnacious leadership of the sharp-tongued Leon Ducasse, it had achieved a position of some influence in the Natal commercial world. As in the Cape there was little love lost between nurserymen and the botanic gardens' staff, who were seen as operating in unfair competition with commercial nurseries because the gardens received government sponsorship in the guise of annual grants.[52] Many Natal nurseries had links with similar establishments in the Cape and Europe, though trade tended to be one way, with garden plants, trees and crop seed being imported from Europe. As a generalisation European nurseries were not keen on introducing new varieties of Natal plants unless, like one of the lilies, they were particularly spectacular.[53] Occasionally they sent their own collectors out to Natal. In the mid-1870s Christopher Mudd, the son of William Mudd, curator of the Cambridge botanic gardens, collected for Messrs Veitch of Chelsea and sent them a new species which was named *Eucomis bicolor*. An interesting example of trade between Natal and Europe was the growing and export to Europe of the Bermudan *Lilium longiflorum* var. *harrisii* which arose because in the late 1890s it was found that Natal-grown bulbs flowered in Europe as late as September, three months after the Bermudan bulbs had ceased.[54]

A few nurserymen in Natal maintained links with Kew, but on a small scale. Between 1884 and 1888, for example, Natal nurserymen sent Kew 17 consignments of plants, seeds and tubers contained in three wardian cases and a number of boxes and parcels. The contents included one cycad (*Encephalartos ghellinckii*); a specimen of 'Kafir potato' (*Coleus* sp.); 'hemp'; two tree species, the broad-leaved Erythrina (*Erythrina latissima*) and a wild pear (*Dombeya rotundifolia*); as well as the following which were believed suitable for European gardens or for greenhouses: *Aloe* sp. *Cycnium racemosum*, *Dolichos* sp.; *Ipomoea* sp.; *Jatropha natalensis*, *Pachycarpus campanulatus* (*Gomphocarpus campanulatus*), *Raphionacme* sp., *Sandersonia aurantiaca* and *Sesamum* sp.

In the same eight years Kew sent Natal nurserymen 35 consignments, of which one was a wardian case containing '40 stove plants and a few economics' and the remaining 34 despatches were entirely made up of packets of seed. These included the seed of such palms as *Euterpe* sp.; the fish tail palm, *Caryota cumingii*; the oil palm, *Elaeis guineensis*; a swamp date palm, *Phoenix paludosa*; and a palmetto palm, *Sabal* sp.. In addition Kew sent seed of such trees as the *Acacia gummifera* from Guinea; the West Indian cedar or Indian mahogany, *Cedrela toona* (*Cedrela adorata*); and the peacock flower flamboyant, *Delonix regia*. Garden plants and grasses in these consignments included a collection of bromeliads, *Eragrostis tef*, *Exbucklandia populnea* and *Zizania aquatica*. It also despatched to Natal nurserymen several consignments of Havana tobacco seed.[55]

THE DEVELOPMENT AND expansion of the nursery industry proved to be one of the major factors in the decline of botanical institutions in late Victorian

---

51 K.A.: Keit, 24 February 1877.

52 See K.C.A.L.: Durban Botanic Society, KCM 43066, Memorandum book 2; and *Natal Government Gazette*, (L955), 24 May 1898.

53 See K.A.: Wood, 18 October 1898.

54 *Kew Bulletin*, (1897), p.406.

55 K.A.: Botanic gardens and nurserymen exchanges volume, 1884 on, Africa, ff147-8.

South Africa. As most botanic gardens by the 1890s had deteriorated into public parks or thinly disguised nurseries, their *raison d'être* as botanic gardens no longer existed. The vital role they had played in introducing plant species of possible economic value often, and in the case of Natal usually, supplied by Kew was no longer necessary. Nor in the eyes of many of the new generation was it desirable. The Natal plant craze declined rapidly in the difficult years of the late 1890s. With nurseries supplying exotic saplings and conventional garden fare and popular interest in botany waning, the link with Kew was inevitably weakened to the point of minimal contact. When it was later renewed, it would be on a very different and professional footing. That this link, an interesting facet of colonial social history, should have splintered and broken in the twilight of the Victorian era is unfortunate. The Natal-Kew link had not only served to present to the British public a positive aspect of a troublesome colony but it had also inspired Natalians to take a pride in their Garden Colony.

Select bibliography

Note on sources

This study of the relationship between Victorian Natal and Kew has as its foundation manuscript material housed in the Kew Archives and the Killie Campbell Africana library. Though both repositories contain much valuable material, one other source has been denied us. In September 1901 Medley Wood, writing to the director of Kew, observed, 'When I took charge [of the Durban botanic gardens] in 1882, I found scarcely any letters and none from Kew'. In the intervening 85 years none of these has come to light elsewhere. The Kew archive does not possess copies of letters sent out to Africa in this period.

Though the completeness of our study has been hindered by the scarceness of surviving correspondence emanating from Kew, fortunately the correspondence received from the colonies is meticulously preserved in the Kew archive in bound volumes. The letters are split into various chronological periods and arranged alphabetically according to the writer. Such volumes occasionally contain Kew memoranda. These and those letters with notes scribbled on them by Kew staff helped us greatly. Equally valuable were the lists of plants and seeds received at Kew and despatched by them, contained respectively in the inward exchange and outward exchange books.

Archival Material

I. *Don Africana library, Durban Municipality*

Curator's annual reports for the Durban botanic gardens, including the early years, 1852, 1855, 1864 and 1865.

II. *Durban city estates department*

File TC15/5J636C contains papers relating to the Durban botanic gardens including the original 1854 title deeds for the land occupied by the Natal Agricultural and Horticultural Society.

III. *Durban local history museum*

1. Two minute books of the Natal Agricultural and Horticultural Society: 1848-1854 (H1130A) and 1855-1864 (H1130B).

2. Miscellaneous papers and photographs relating to the Durban botanic gardens.

IV. *Forestry archive, Eshowe and Pietermaritzburg*

Miscellaneous reports relating to forestry in Natal and Zululand.

V. *Killie Campbell Africana Library, Durban*

1. Durban Botanic Society papers: KCM 43065-43074, including two memorandum books, an herbarium visitors' book and two tea room visitors' books. The library also has a collection of printed curator's annual reports.

2. Pietermaritzburg Botanic Society: This collection includes a number of curators annual reports, a cash book and journal of expenses, a collection of curators' diaries from 1891, photographs and some miscellaneous papers.

3. Natal-Kew correspondence: Copies of letters sent from Natal, mainly concerned with the Durban Botanic Society; transcribed copies presented by R.G. Strey.

VI *Keit papers*

Letters from Wilhelm Keit to his family in Dresden. Retained by the Keit family in South Africa.

VII *Archives of the Royal Botanic Gardens, Kew*

A. *Bound volumes of letters used:*

1. English letters, 1849, vol XXVII.
2. English letters, 1853, vol XXXIII.
3. English letters, 1854, vol XXIV.
4. African letters, 1844-58, vol LIX.
5. African letters, 1859-65, vol LX.
6. South American letters, 1841-51, vol LXX.
7. South African letters, 1865-1900, A- G, vol 189.
8. South African letters, 1865-1900, H-N, vol 190.
9. South African letters, 1865-1900, P- W, vol 191.
10. Eastern Cape Colony and Natal letters, 1901-15, vol 194.

B. *Inward exchange books* (plants and seeds received by Kew), (Kewensia rooms) 1848-58, 1859-67, 1868-87, 1888-92, 1892-99 and 1899-1908.

C. *Outward exchange books* (plants and seeds despatched by Kew), (Kewensia rooms),1848-59, 1860-9, 1869-81, 1881-95 and 1896-1923.

D. *Plant list volumes*
For the early Natal period see vol I, Australia, New Zealand and Polynesia, 1845-63, and vol II, 'Orient, Africa and America, 1852.

E. Botanic gardens and nurserymen exchanges volume (1884 on), (Kewensia rooms).

F. Colonial floras volume (Kewensia rooms).

G. Miscellaneous reports: Natal Botanic Gardens, 1866-1909, vol 305.

H. Natal miscellaneous, 1862-1909, vol 305.

I. South Africa: Botanists, 1859-96, Miscellaneous reports, vol 12.6.

VIII *Natal archival depôt*

1. The colonial secretary's office papers contain copies of some of the unpublished curator's annual reports for the Durban botanic gardens.

2. Papers relating to forestry in Natal are contained in the government house papers.

3. The archive possesses a copy of *Correspondence and reports relative to the state of botanical enterprise in Natal, 1882*, (Pietermaritzburg, 1884).

IX *Natal herbarium, Durban*

1. Newscutting books and curator's annual reports for the Durban botanic gardens.

2. Notes on curators compiled by R.G. Strey.

3. 'Kew lists, from 1876 to 1906, Natal Government Herbarium'. A typed volume of plant names identified by the Kew herbarium with frequent comments on individual species.

4. The herbarium library also contains a set of the *Kew Bulletin* and *Index Kewensis* as well as an incomplete set of *Curtis's Botanical Magazine, Icones Plantarum* and the *Journal of Botany*.

X *National botanic gardens of Ireland, Glasnevin*
Correspondence relating to Wilhelm Keit.

Newspapers and periodical publications

Agricultural Journal (Natal)
Bothalia
Curtis's Botanical Magazine
Daily News (Natal)
Gardeners' Chronicle
Genera Plantarum, (G. Bentham and J.D. Hooker, London, 1862-81).
Grevillea
Hooker's London Journal of Botany (Kew, 1842-57)
Icones Plantarum (1837 on)
Illustrated London News
Index Kewensis (1885 on)
Journal of Botany
Journal of the Royal Horticultural Society
Natal Colonist
Natal Government Gazette
Natal Mercury
Natal Wildlife
Natal Witness
Royal Gardens, Kew: Bulletin of Miscellaneous Information (1887-1941)
Report on the progress and condition of the Royal Gardens at Kew, (Sir William Hooker, 1856-65; Sir Joseph Hooker, 1866-82. No reports published, 1883-86). *South African Journal of Science Transactions of the South African Philosophical Society*

Specialised Printed Works

Mea Allan, *The Hookers of Kew, 1785-1911*, (London, 1967).

D.E. Allen, *The Victorian fern craze: A history of pteridomania*, (London, 1969).

E. Armitage, 'Some observations on the botany of Natal', in J. Chapman, *Travels in the interior of South Africa*, (London, 1868), 2.458-65.

Mary Elizabeth Barber, 'Wanderings in South Africa by sea and land, 1879', *Quarterly Bulletin of the South African Library*, 18, 1, (1963), 6-11.

A. Batten and H. Bokelmann, *Wild flowers of the Eastern Cape Province*, (Cape Town, 1966).

A.W. Bayer, 'Aspects of Natal's botanical history', *South African Journal of Science*, vol 67, no. 8, (August 1971),

401-11. 'Discovering the Natal flora', *Natalia*, 4, (December 1974), 42-8. Introduction to *Flower paintings of Katharine Saunders*, (Tongaat, 1979).

W.J. Bean, *The Royal Botanic Gardens, Kew: Historical and descriptive*, (London, 1908).

J.W. Bews, *An introduction to the flora of Natal and Zululand*, (Pietermaritzburg, 1921).

M. Bingham, *The making of Kew*, (London, 1975).

W. Blunt, *In for a penny: A prospect of Kew Gardens*, (London, 1978).

'Botanic Garden, Durban', *Kew Bulletin*, (1900), 12-15.

'The botanical aspect of Natal', *Natal Journal*, 2, (1858), 175-87.

A.M. Bottomly, 'An account of Natal fungi collected by J. Medley Wood', *South African Journal of Science*, 13, (1916), 424-46.

D. Brandis, 'Forestry in Natal', *Kew Bulletin*, (1895), D1-5.

James Britten and G.S. Boulger, 'Biographical index of British and Irish botanists', *Journal of Botany*, vols 26-9, (1888-91).

Lucille H. Brockway, *Science and colonial expansion: The role of the British Royal Botanic Society*, (London, 1979).

E.H. Brookes and C. de B. Webb, *A history of Natal*, (University of Natal Press, 1965).

A.C. Brown (ed), *A history of scientific endeavour in South Africa*, (Cape Town, 1977).

N.E. Brown, 'John Medley Wood, 1827-1915', *Kew Bulletin*, (1915), 417-9.

Sir Charles Bruce, *The broad stone of empire*, (London, 1910), 2 vols.

Rev. John Buchanan, *Revised list of the ferns of Natal*, (1875).

'Centenary of the Royal Botanic Gardens, Kew', *Kew Bulletin*, (1941), 201-9.

Daphne Child, *A merchant family in early Natal: Diaries and letters of Joseph and Marianne Churchill*, (Cape Town, 1979).

Alice M. Coats, *Quest for plants: A history of horticultural explorers*, (London, 1969).

L.E. Codd, B. de Winter and H.B. Rycroft (et al), *Flora of Southern Africa*, (Pretoria, 1966).

Correspondence and reports relative to the state of botanical enterprise in Natal, 1882, (Pietermaritzburg, 1884), (copies in Natal Archives, Pietermaritzburg, and K.C.A.L.).

M.C. Cooke and C. Kalchbrenner, 'Natal fungi collected by J.M. Wood, Inanda', *Grevillea*, X (1881), 26-7.

R. Desmond, *Dictionary of British and Irish botanists and horticulturalists*, (London, 1977).

'Historical account of Kew to 1841', (Thiselton Dyer), *Kew Bulletin*, (1891), 279-327.

'Emigration to Natal', *Illustrated London News*, 16 March 1850.

'Encephalartos woodii', *Kew Bulletin*, (1914), 250-1.

Maurice S. Evans, *The fertilisation of flowers*, (Durban c. 1894). *Pollen: A paper read at a meeting of the Natal Microscopical Society, 18 November 1878*, (Durban, 1878).

Barbara Evegard and Brian D. Morley, *Wild flowers of the world*, (London, 1970).

A short account of the fern collections, Royal Botanic Gardens, Kew, (leaflet, 1976).

H.G. Fourcade, *Report on the Natal forests*, (Natal government report, Pietermaritzburg, 1889).

William Gardener, 'The Royal Botanic Gardens at Kew: The work of Sir William Jackson Hooker and John Lindley', *History Today*, XV, 10, (October 1965), 688-96.

Janet M. Gibson, *Wild flowers of Natal*, (Durban, 1975 and 1978), 2 vols.

J.R. Green, *A history of botany in the United Kingdom*, (London, 1914).

Bill Guest and John M. Sellers, *Enterprise and exploitation in a Victorian colony: Aspects of the economic and social history of colonial Natal*, (University of Natal Press, 1985).

Mary Gunn and L.E. Codd, *Botanical exploration of southern Africa*, (Cape Town, 1981). 'Additional biographical notes on plant collectors in southern Africa', *Bothalia*, 15,3 and 4, (1985), 631-54.

Miles Hadfield, *A history of British gardening*, (London, 1969).

Alan F. Hattersley, *The British settlement of Natal*, (Cambridge University Press, 1950). *The Natalians*, (Pietermaritzburg, 1940).

W.F. Harvey, *The genera of South African plants*, (London, 1868 edition by J.D. Hooker). Thesaurus capensis, (Dublin, 1859), 2 vols and O.W. Sonder, Flora capensis, (?London, 1859-66), vols I-III.

Ellison Hawks, Pioneers of plant study, London, 1928).

M. Henderson and Johan G. Anderson, *Common weeds of South Africa*, (Botanical survey memoir no. 37, (Pretoria 1966).

W.P.M. Henderson, *Durban: Fifty years' municipal history*, (Durban 1904).

F.N. Hepper (ed), *Royal Botanic Gardens, Kew: Gardens for science and pleasure*, (H.M.S.O., 1982).

Agnes H. Hicks, *The story of the forestal*, (London, 1956).

Rev. William C. Holden, *History of the Colony of Natal, South Africa*, London, 1855).

J.D. Hooker, *Himalayan journals,* (London,1854), 2 vols.

Sir Joseph D. Hooker (ed), *Journal of the Rt Hon Sir Joseph Banks*, (London, 1896).

N. Hurwitz, *Agriculture in Natal, 1860-1950*, (Oxford University Press, 1957), vol 12.

L. Huxley, *Life and letters of Sir J.D. Hooker, based on materials collected and arranged by Lady Hooker*, (London, 1918), 2 vols.

B.D. Jackson, *Guide to the literature of botany*, (London, 1881).'List of the collectors whose plants are in the herbarium of the Royal Botanic Gardens, Kew, to 31 December 1899', *Kew Bulletin*, (1901) 1-80.

Johnson's Gardeners' Dictionary, (C.H. Wright and D. Dewar), (London, 1894 edition: 1910 reprint).

M.C. Karsten, 'J. Medley Wood, in *'Dictionary of South African Biography*, II.857.

H Ronald King, *The world of Kew*, (London, 1976).

B.J.T. Leverton, *Records of Natal, 1823-1828*, (Pretoria, 1984).

Conrad Lighton, *Cape floral kingdom*, (Cape Town, 1960).

Leslie Linder (ed), *The journal of Beatrix Potter*, (London, 1966).

A list of the collectors whose plants are in the herbarium of the Royal Botanic Gardens, Kew, to 31 December 1899, *Kew Bulletin*, (1901), 1-72.

Donal P. McCracken, 'The development of botanic institutions in nineteenth century Natal and West Africa', *Journal of the University of Durban-Westville*, new series 1, (1984), 107-15.

'Durban botanic gardens, 1853-1913', *Garden History*, 15.1, (1987), 64-73.

'The indigenous forests of colonial Natal and Zululand', *Natalia*, 16, (1986), 19-38.

'William Keit and the Durban botanic gardens', *Bothalia*, 16, (1986), 71-75.

'Natal's botanic gardens: their role in the colonial economy and society', paper presented to the conference of the South African Historical Society, University of Cape Town, January 1985.

Donal P. McCracken and Eileen M. McCracken, *The way to Kirstenbosch*, (N.B.G., 1988).

Eileen M. McCracken, *The palm house and botanic gardens, Belfast*, (Belfast, 1971).

M.J. McKen, *The ferns of Natal*, (Pietermaritzburg, 1869). McKen, obituaries and notes on, *Journal of Botany*, (1872), 223; *Natal Mercury*, 25 April 1872, 8 May 1872 and 10 July 1954.

Christopher Merrett, 'William Stanger and the early years of cartography in Natal, 1845-54', *Natalia*, 9, (December 1979), 30-5.

W.F. Monypenny and G.E. Buckle, *The life of Benjamin Disraeli*, 6 vols, London, 1910-20).

T. Mower Martin and A.R. Hope Moncrieff, *Kew Gardens*, (London, 1908).

'Natal', *Kew Bulletin*, (September 1887), 12-3.

'Natal plants', *Journal of Botany*, 38, (1900), 192-3.

'Natal tea', *Kew Bulletin*, (1888), 87-8.

'National Botanic Gardens of South Africa', *Kew Bulletin*, (1913), 309.

E. Charles Nelson and Eileen M. McCracken, *The brightest jewel: A history of the National Botanic Gardens of Glasnevin, Dublin*, 1 (Kilkenny, 1987) and

Wendy Walsh, *An Irish flower garden*, (Kilkenny, 1984).

Marianne North, (ed Mrs J.A. Symonds), *Recollections of a happy life*, (London, 1892), 2 vols.

F.A. Novak, (ed J.G. Barton), *The pictorial encyclopædia of plants and flowers*, (London, 1966).

D. Oliver et al, *Flora of tropical Africa*, (London, 1868-1937).

The Orangery: Royal Botanic Gardens: Kew, (HMSO, c. 1971).

Roy Osborne, 'Encephalartos woodii' *Encephalartos*, 5, (1986), 4-10.

Keith Coates Palgrave, *Trees of southern Africa*, (Cape Town, 2nd edition, 1983).

Paxton's Botanical Dictionary, (London, 1868 edition).

E. Percy Phillips, 'A brief historical sketch of the development of botanical science in South Africa and the contribution of South Africa to botany', *South African Journal of Science*, XXVII, (November, 1930), 39-80.

The genera of South African flowering plants, (Botanical survey memoir, no. 25, (Pretoria, 1951).

Kristo Pienaar, *The South African What flower is that?*, (Cape Town, 1984).

R.W. Plant, 'Notice of an excursion in the Zulu country', *Hooker's Journal of Botany*, IV, (1852), 257-65; and note on R.W. Plant, *Hooker's Journal of Botany*, IV, (1852), 222-3.

I.B. Pole Evans *et al*, *Flowering plants of South Africa*, (London and South Africa, 1921).

Adrian Preston, *The South African diary of Sir Garnet Wolseley*, (Cape Town, 1971).

Wyn Rees (ed), *Colenso letters from Natal*, (Pietermaritzburg, 1958).

J.H. Ross, *Flora of Natal*, Botanical survey memoir no. 39, (Pretoria, 1972).

John Sanderson, 'Rough notes on the botany of Natal', in J.C. Chapman, *Travels in the interior of South Africa*, London, 1868), II. 443-58.

B.D. Schrire, 'Centenary of the Natal Herbarium, 1882-1982', *Bothalia*, 14, 2, (1983), 223-36.

S.P. Sherry, 'History of the wattle industry in Natal', *Natalia*, 3, (September 1973), 40-4.

T.R. Sim, *Tree planting in Natal*, (Pietermaritzburg, 1907).

C.A. Smith, *Common names of South African plants*, Botanical survey memoir no. 35, (Pretoria, 1966).

Rudolph G. Strey, 'The father of Natal botany: John Medley Wood', *Natalia*, 7 (December 1977), 43-5.

'The second curator of the botanic gardens, Durban, arrived 100 years ago', *Natal Wildlife*, (November 1972), p.18.

P.C. Sutherland, *Journal of a voyage in Baffin's Bay and Barrow Straits*, (London, 1852), 2 vols.

Ernest R. Thorp, *Durban Botanical Gardens*, (Durban 1970).

W.B. Turrill, *Joseph Dalton Hooker: Botanist, explorer and administrator*, (London, 1963). 'Royal Botanic Gardens, Kew', *Journal of the Royal Horticultural Society*, 84.6, (1959),N256-64. *The Royal Botanic gardens: Kew: Past and present*, (London, 1959).

A vision of Eden: The life and work of Marianne North, (London, 1980).

K.G.T. Watson, *Tongaati: An African experience*, (London, 1960).

Alain White and Boyd L. Sloane, *The stapeliae* vol 1, (California, 1933).

B.S. Williams, *Select ferns and lycopods*, (London, 1873).

John Medley Wood, *An analytical key to the natural orders and genera of Natal indigenous plants*, (Durban, 1888). 'Botanic gardens', in *Twentieth century impressions of Natal*, (London, 1900). *Classification of ferns*, (Durban, 1879).

The flora of Durban and vicinity, (Durban, 1887).
A guide to the Natal Botanic Gardens, (Durban, 1883).
Guide to the trees and shrubs of the Natal Botanic Gardens, (Durban, 1897).

List of trees, shrubs and a selection of herbaceous plants growing in the Durban Municipal Botanic Gardens, (Durban, 1915).

Natal Plants, (Durban, 1899-1912), vols 1-6, (vol 1 with Maurice S. Evans).

'Native herbs', in *Natal directory and almanac*, (Pietermaritzburg, 1896), 260-5.

A popular description of Natal ferns designed for the use of amateurs, (Durban, 1877).

A preliminary catalogue of indigenous Natal plants, (Durban, 1894).

Report on the Natal Botanic Gardens and Colonial Herbarium, (1896-1910).

'Revised list of the flora of Natal', *Transactions of the South African Philosophical Society*, XVII.2, (1908), 121-280.

'A chat with J. Medley Wood, ALS', *Agricultural Journal*, VI.II, (26 June 1903), 345-7.

'J.M. Wood', *Journal of the Linnean Society*, XXV, (1890), 171.

'Medley Wood', in Alain White and Boyd L. Sloane, *The stapeliae*, (California, 1933), I.120.

Index

Page numbers to the illustrations are printed in italics

A

A guide to the Natal Botanic Gardens 58
A popular description of the Natal ferns 41
A revised list of the ferns of Natal 41
Aberdeen, The earl of, xvii
Acacia 44, 70
   gummifera 84, 86
   mearnsii 44, 54
   pycnantha 84
Addison, W. 53
Adelaide 45
Adenium 73
Adlam, R.W. 65
Aiken, J. 57
Aiton, W.T. xvi, xvii
Alberta magna 18, *81*, 81
Albuca gardeni 14
   nelsonii 32, 32
Alfred County 33, 78
Allison, A. 71, 72
Allison M. 71, 72
Aloe 86
   arborescens 81
   bainesii 43, 43, 63, *81*, 81
   cooperi 81
   kraussii 81
Alsophila capensis 81
America 9
American revolution xvii
Amomum cardomomum 19
Anemone fanninii 47
Anglo-Boer war, The first, xxi
Anglo-Boer war 55, 76
Anglo-Zulu war xxi, 55, 74
Angraecum 47
   maudiae 71
Ansellia 47
Antigonon leptopus 45
Araucaria 84
   araucana 84
   bidwillii 85
   heterophylla 54, 85
   cunninghamii 85

Argyrolobium ascendens 2
   sandersonii 15
   sutherlandii 17
Armitage, Rev.E. 8, 15
Arundessaria falconeri 84
Asaba 62
Asclepiadaceae sp. 24
Attwood, S. 71
Augusta, Dowager Princess, xvi
Augusta, Princess, xv, xvii
Australia xix, xxi, 9
Ayrton, A.S. 28
Azalea 19

B

Baikie's Niger expedition 11
Baines, T. 43
Baker, J.G. 4, 29, 32, 60, 73
Baltic timber 18
Bambatha insurrection 76
Banks, Sir J. xvii, 27
Barber, Mrs M.E. 33, 71
Barberton daisy 65
Barker, J.G. 82
Barkly, Sir H. 10, 63
Barrow, Sir G. 16
Barter, C. 11
Basananthe sandersonii 15
Basle 50
Bedford, The duke of, xvii
Begonia geranioides 25
   natalensis 14
   sutherlandii 17, 25
Bell, W. 17
Beningfield, S. 26, 64
Bentham, Sir G. 5, 9, 30, 60
Bignonia picta 85
Bishopstowe 32
Black, A.A. 29
Black wattle 54
Blue gums 54
Boehmeria nivea 53
Boers xx, xxi
Bolus, H. 23, 31, 53, 71, 73, 74, 82
Boswellia serrata 84
Botanic Gardens Road 58
Botanic stations 62
Bourbon 52
Bowie, J. 2

Bowker, Col J.H. 33, 41, 56, 70, 71, 79, 82, 83
Brachycorythis 47
Brisbane xix
Brazil 84
Brickhill, J. 8, 62
Britain 22, 23
British Guiana 9
British Guiana botanic gardens 71
Brown, N.E. 11, 29, 42, 73, 76
Brown, Rev Dr J.C. 31, 63
Browne, Napier 41
Brownleea 47
Brunsfelsia calycina 19
   hydrangeaeformis 19
Brussels 50
Buchanan Rev. J. 41, 45
Buchanan Rev. S. 67
Bulbophyllum 47
Bulwer, Sir H. 54, 58
Bushman's River 73
Button, Edward 45
Buttonia 45
   natalensis 81
Byrne 23
Byrne, J. 2
Byrne settlers 3

C

Calanthe natalensis 63
Calcutta 77
Cambridge botanic gardens 86
Camellia 19
Canada 16
Cane smut 52
Capability Brown xvi
Cape Colony xx
Cape of Good Hope xvii
Cape Town 21, 22, 23, 45, 62, 77, 78
Caralluma lutea 15
Carapa guianensis 26
Carissa 24
Caroline, Queen, xv, xvi
Caryota cumingii 86
Cedrela adorata 86
   toona 86
Cedrus deodara 85
Cephaelis ipecacuanha 26
Cereus macdonaldiae 19

Ceropegia sandersonii 15
Ceylon 9, 45, 52, 53, 60
Chamberlain, Joseph 61
Chambers, Sir W. xvi
Charlestown 78
Cheadle 5
Cheilanthes capensis 81
　　hirta 81
Chelsea physic garden xvii, 5
China 11
China grass 53
Chinchona 53
　　succirubra 53
Chlorocoden whitei 33
Chrysanthemum 26
Cissus 24
　　fragilis 24
　　sandersonii 15
Clarence, A. 17
Clerodendron 24
　　myricoides 24
Clivia gardenii 14
Coffea stenophylla 84
Coffee 52, 69
Cola acuminata 84
Colenso 76
Colenso, Bishop J.W. *18*, 19, 23, 26, 32, 49
Colenso, Miss 83
Colenso, Mrs F. 19
Coleus 83 sp. 86
Colley, Gen.,Sir G.P. 33
Collins, W.M. 17
Colonial exhibition 8
Colonial floras 9
Colonial herbarium *68*
Colonial office 16, 58, 61
Colonial surveyor-general 12
Colonial treasurer 9
Commissioner of woods & forests xvi
Cooper, T. 11
Copernicia cerifera 84
Corycium 47
Cotton 3
Crinum 19, 24
Crystal Palace 12
Currie, Anna L. 56
Currie, Henry William 56
Curtis's Botanical Magazine 9, 14, 30, *31*, 62
Cycnium racemosum 86
Cyperus 52
Cyrtorchis 47

D
Dahlia 19
Dallimore, W. 64
Dargle 47
Darwin, C. 19, 28
Davies, G.H. 74
Delonix regia 86
Dingaan 73
Dioscoria crinita 63
Disa 47
　　cooperi *48*
Disperis 47
Disraeli, Benjamin Lord xx, 28
Dolichos 86
Dombeya burgessiae 24
　　rotundifolia 86
Drakensberg xx, 53, 55, 73, 78
Dresden 50
Drège, W.F. 2
Dublin 10, 49, 73
Ducasse, L. 86
Dunn, John 74
Durban xx, xxi, 4, 6, 14, 18, 20, 21, 23, 43, 46, *57*, 62, 76, 78
Durban botanic gardens xxi, 4, 10, 22, 17, 23, 32, 44, 47, 49, 50, 52, 53, 54, *59*, 65, 66, 70, 76, 77, 80, 84
Durban Botanic Society 58, 59, 62, 65
Dutch East India Company garden xxi, 27
Dutch House xvi
D'Urban, W.S.M. 17

E
East Griqualand 80
Eastern Cape 73, 74
East-South-Central-African Herbarium (E.S.C.A.H.) 72
Ekukanyeni 19
Elaeis guineensis 86
Elephantorrhiza burchelli 62
　　elephantina 62
Elliot, George 49
Emily 3
Empress Eugenie 33
Encephalartos 45, 80
　　barterisis 81
　　brachyphyllus 81
　　caffer 81
　　ghellinckii 81, 86
　　longifolius xviii
　　mackenii 4
　　villosus 4, 4, 81
　　woodii 74, 81
Eragostis abyssinica 84
　　tef 86
Erica alopecurus 17
Eriospermum mackenii 4
Erythrina latissima 86
Escallonia macrantha 19
Eshowe 74
Eucalyptus 54, 69
　　globulus 54
Eucomis bicolor 63, *63*, 81, 86
Eulophia revoluta 47
Eulophidium mackenii 4
Euphorbia 79 sp. 24
Euterpe sp. 86
Evans, M.S. 70, *71*, 71, 73, 83
Exbucklandia populnea 85, 86
Exchange books 20

F
Fannin, G.T. 15 J.F. 47
Fernando Po 82
Ferns of Natal 45
Fielden, H.W. 55
Field's Hill 14
Fitch, W. 9, 28
Flora Capensis 9, 11, 15, 24, 31, 63, 70, 71, 85
Flora Indica 9
Flora Natalensis 7, 9
Flora of Natal 71
Forests 54
Fort Beaufort xxi
Fortune, R. 23
Foul Point 11
Frederick, Prince of Wales xv
Frere, Sir B. 33
Fundisweni 33

G
Garden, R. 14, 26, 64
Gardeners' Chronicle 65, 76, 85
Gardenia 19
　　fortunei 85
　　intermis 85
Genera of South African Plants 9, 31
Genera Plantarum 5, 30, 60

George II xv
George III xv, xvi, xvii
George IV xvii
Geranium incanum 62
Gerbera jamesonii 65
Gerrard, W.T. 3, 4, 9, 10, 11, 20, 27, 45, 47, 67, 73
Gerrardanthus 10, 41, 79
Gerrardina 10
 foliosa 10
Gladiolus 24
 oppositiflorus 80
 salmoneus 81
Glasgow university xvii
Glasnevin 50
Glasnevin botanic gardens 14
Gloriosa 19
 plantii 8, 24
Godwin, Dr G. 16
Gomphocarpus campanulatus 86
Graaff-Reinet xxi, 62
Grahamstown xxi, 62
Grahamstown botanic gardens 44
Great palm house 18
Great stove, The, xvi
Greenacre, Sir B. 68
Greenland 16
Grevillea 85
Grewia caffra 33
Greyia sutherlandii 17, 33
Gueinzius, W. 2, 14
Guttaria caffra 24

H
Habenaria 40, 47
Haemanthus insignis 21
 katharinae 22, 40, 40, 50, 51
 natalensis 14, 24
Halimum sp. 24
Halleria lucida 52
Haloxylon ammodendron 84
Harpephyllum caffrum 62
Harvey, J. 31
Harvey, W. 2, 9, 10, 11, 20, 27, 40, 63, 73
Haygarth, W. 41, 70
Hebeclinium 19
Heliophila scandens 62
Hely-Hutchinson, Sir W. 63, 67, 74
Hemiteia 69
 woodii 69

96

Hesperaloe engolmannis 84
Hevea brasiliensis 53
Hibiscus 19
 natalensis 82
 pedunculatus 85
 physaloides 82
 schizopetalus 85
 vitifolius 82
Hilton College 46, 85
Hindson, W.R. 85
Hislop, A. 66
Hobart xix
Hollow, The, xvi
Holley Walk xv, xvi
Honduras 9
Hong Kong 9
Hooker, Sir J. 3, 5, 9, 11, 19, 23, 27, 28, 29, 32, 33, 40, 41, 44, 45, 46, 47, 49, 53, 55, 56, 57, 64, 79, 80, 82, 85
Hooker, Sir W.J. xvii, xix, 2, 3, 5, 6, 8, 9, 10, 14, 16, 18, 25, 26, 30, 31, 58
Hooker's Journal of Botany 6
Hoya 24
Huernia 82
 hystrix 82
Hulett, Sir J. 43
Huttonaeo 47
Hydrocotyle var. *monticola* 82
Hyphaene natalensis 81
Hypoxis 14
 elata 25
 latifolia 25

I
Icones Plantarum 11, 12, 30, 42, 85
Illustrated London News 2
Inanda 41, 55
Index Floral Sinensis 85
Index Kewensis 12, 60
India xxi, 9, 27, 85
Indian Ocean 6
Ipomoea 86
 gerrardii 25
Ireland xx, 9
Isandhlwana 55, 56, 73
Itoli mountains 12

J
Jacaranda mimosifolia 84
Jackson, D.D. 60 J.R. 83
Jamaica 3, 60
Jambosa vulgaris 26

Jameson, R. 41, 65
Jardin des Plantes xvii
Jasmine 6
Jasminum sp. 24
Jatropha natalensis 86
Jenkinson, H 74
Johannesburg 78
Johnson's Gardeners' Dictionary 80
Journal of Botany 46, 70
Jubilee conservatory 67, 68, 75

K
Karkloof 53
Keate, R.W. 14, 63
Keit, W. 32, 49, 50, 52, 53, 56, 58, 65, 86
Kenya 51
Kew xv, 2, 45
Kew Bulletin 62, 71, 75, 76, 77, 85
Kew herbarium xvi, 9, 28, 42
Kew House xvi, xvii
Kew House Estate xv
Kew Jodrell laboratory 29
Kew museum, xxi, 18, 82
Kew Palace xvi
Kew Report 11, 30, 50, 52, 53, 63
Khoisan xxi
Kimberley xxi
Kimberley, Lord 58
King William's Town xxi, 62
Kirstenbosch xxi, 77
Kleinia fulgens 25
Kniphofia modesta 71
Kokstad 81
Krauss, F. 2

L
Ladysmith 78
Laidley and Company 72
Large, Sam 62
Lastrea plantii 8
Laurel 6
Lauth, F. 70
Lebombo 74
Leipzig 2
Licuala grandis 84
Lilium longiflorum var. *harrisii* 86
Limonium peregrinum 2
Lindley, Dr J. xvii, 27
Lindley report xvii
Linnean Society 5, 71
Linum thunbergia 16
Liparis 47
Lissochilus 47

Lissochilus sandersonii 15
Littonia modesta 51, 51, 81
Liverpool 19, 45
Lloyd, C.B. 71
Lobelia 82
Lomaria 14
   cycadioides 81
   zamaefolia 81
   zamioides 81
London xx, 4, 6, 18, 32
Love Lane xvi
Lyell, Mrs K. 19

M
Mackenia 4
MacOwan, P. 70, 73, 82
Madagascar 11
Madeira 41
Magaliesberg 14
Magnolia 26
Mailship 78, 86
Malaria 73
Malmesbury, Lord, xx
Mann, Dr R.J. 51
Maputaland 73, 75
Maquaira grandiflora 85
Marabastad 45
Marriott, W.E. 66
Mashonaland, 43
Masson, F. xvii, xviii, 11
Mauritius 9, 10, 11, 30, 45, 52
McGillvary, M. 45
McKen, M. 3, 3, 4, 6, 10, 14, 16, 20, 21, 23, 26, 27, 40, 41, 43, 44, 45, 46, 52, 64, 67, 70, 73, 76, 79, 80, 82, 84
McKenzie, Mrs 73
Meade J.L. 15, 58 R.H. 58
Melanea 47
Melbourne xix, 45
Melhania erythroxylon 84 sp. 84
Meller, Dr C. 45
Merlin's Cave xv
Methonica plantii 24
Mexico 45
Mhlatuze 73
Microscope 68
Miller, P. xvii
Millettia sutherlandii 17
Mitchell, G. 65
Mkuzi 74
Mondia whitei 33
Monodenia 47

Moore, Dr D. 14, 50
Moreton Bay Pine 85
Morris, Sir D. 60
Mpande 6
Mr Bull's Establishment for New and Rare Plants 32
Mudd, C. 86 W. 86
Musa ensete 84
Mystacidium 47
Myurus inaequalis 24
Mzimvubu river 18

N
Namaqualand 80
Natal 19, 77
Natal Agricultural and Horticultural Society 2, 44, 57
Natal Colonist 14, 49
Natal earthworms 71
Natal flame bush 81
Natal Government Railway 21
Natal Herbarium 76
Natal Mercury 6, 23
Natal Natural History Association 49
Natal Plants 70, 73, 79
Natal Times 14
Natal Witness 85
Natalia, Republic of, xx
National Botanic Gardens of South Africa, xxi, 77
Natural History Museum 33
Nelson, W. 32
Nesfield, W.A. xviii
Nettleship, E. 83
New Gardeners' Dictionary 5
New South Wales xxi
New Zealand 9 30
Newberry. W.J. 66
Newfoundland 64
Newnham, Rev. W.A. 85
Ngoye 74
Nguni xx, 74, 83
Nicholson, G. 64
Nkandhla-Qudeni forest region 78
Noodsberg 5 41
Norfolk 55
Norfolk Island pine 54
North, Marianne 32, 33, 47
North, Marianne, The paintings of, *34-39*
North Gallery 32
Nottingham 50
Nursery industry 86
Nymphaea coerulea 72

Nymphaea stellata 72

O
Ochnaceae sp. 24
Oldham, R. 11
Oliver, Prof. D. 9, 11, 12, 29, *30*, 40, 60, 73
Orange Free State 80
Orange River Sovereignty xx, 51
Orangery xvi, xviii, *xx, 25*
Orchidaceae 47
Oregon 3
Ormonde Lodge xv
Owen, Miss M.C. 2, 73
Owen, Prof. R. 28
Ox wagon 5, 21
Oxalis crenata 84
Oxyanthus natalensis 24

P
Pachycarpus campanulatus 86
Pachypodium 74
Pagoda xviii
Palm house, The great, *xv, xx*
Pamplemousses botanic gardens 45
Pappe, Dr L. 31
Paris 50
Parmestera cirifera 84
Passiflora maliformis 85
   racemosa 85
Pavonia multiflora 85
Paxton, Sir J. xvii, xxi
Paxton's Botanical Dictionary 24, 85
Peace, W. 51
Peek, Sir H.W. 71
Peponia mackenii 4
Peradeniya botanic gardens 53
Phillips, E.P. 31
Phoenix paludosa 84, 86
   reclinata 81
Phoenix Park 49
Phytolacca 82
Picrasma excelsa 85
Pietermaritzburg 5, 10, 14, 21, 32, 44, 62, 76, 78
Pietermaritzburg botanic gardens 22, 44, 50, 65, *66,* 77
Pietermaritzburg Botanic Society 12, 65
Pine, Sir B. 16

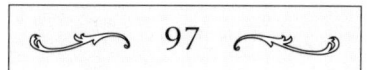

Pineapple lily, 81
Pinus radiata 54
Plant, Mrs 8, 52
Plant, N. 5
Plant, R. 3, 5, 6, 7, 8, 9, 11, 15, 16, 20, 21, 23, 70, 73
Poeppig, G.F. 2
Polystachya, 47
Pondoland 16
Pope, Alexander xv
Port Elizabeth xxi, 62, 72
Port Natal xx, 2
Portuguese East Africa 7, 80
Postal service 21, 22
Potter, Beatrix 60
Prain, D. 77
Pretoria xxi, 76
Protea 83
Pseudophoenix sargentii 85
Pterodiscus speciosus 81
Pyrethrum cinerariaefolium 84

Q
Queensland xxi, 45
Queenstown xxi
Queen's Cottage xvi, 64
Quercus suber 26
Quilimane 80

R
Ralph, J. 71
Raphionacme sp. 86
Rauvolfia caffra 83
Rehmann, A. 41
Report on the Gardens 31
Rhododendron 26, 27
Rhododendron Walk xvi
Rhus sp. 24
Richardia hastata 24
  rehmannii 65
Richmond xv
Richmond Lodge xv, xvi
Richards Bay 73
Rinderpest 76
Robertson, Rev R. 31
Robertson, W.P. 17, 64
Rondeletia speciosa 19
Ross 71
Roupellia grata 26
Royal botanic gardens of Ireland 50

98

Royal Society 28, 75
Royena sp 24
Rubber 53
Rubiaceae 24
Rudatis, A. 76

S
Sabal sp. 86
Sanders and Company 72
Sanderson, J. 5, 7, 12, 14, 15, 16, 21, 22, 23, 26, 28, 41, 46, 47, 49, 50, 58, 63, 64, 65, 73, 76, 78, 79, 82, 83, 85
Sanderson, S. 73
Sandersonia aurantiaca 13, 14, 24, 86
Sarcostemma viminale 83
Satyrium 47
Saunders, C. 40, 71, 74
Saunders, J. 40
Saunders, Mrs K. 4, 32, *33*, 40, 41, 71, 73, 83
Saunders, W.W. 5, 7, 11, 31
Sawmills 53
Schizochilus 47
Schlechter, Dr R. 71
Schönland, S. 73
Schöpflin, F. 74
Scilla 82
Scott, J. 17
Sea Cow Lake 21
Senecio sandersonii 15
Sequoia gigantea 85
Serenoa serrulata 85
Sesamum sp. 86
Seychelles 9, 21, 30
Shanghai 23
Shepstone, Sir T. 66, *67*, 83
Sim, T.R. 76, 86
Smith, J. 23, 27, 30
Smith II, J. 64
Sonder, O.W. 63
South African College 31
Species Filicum 27
Spirostachys africana 83
St Lucia 7, 83
Stanger, Dr W. 12, 16
Stangeria eriopus 6
  paradoxa 6, 6, 14, 24
Stapelia 8
  gigantea 8, 8 sp 24
Statice rytidophylla 2
Stegobium paneceum 79
Stenoglottis 47

Strelitzia 19, 24
Streptocarpus 7
  cooksonii 10
  grandis 14
  polyanthus 10
saundersii 7
Sugar 3
Sugar cane 51, 52, 69
Sutherland 16
Sutherland, Dr P.C. 12, 14, *15*, 15, 16, 17, 21, 23, 28, 46, 49, 51, 54, 65, 68, 73, 82, 83
Swanfield,H. 74
Swaziland 80
Sydney 45
Synopsis Filicum 85
Synopsis Filicum Capensium 10, 45
Synopsis of all known ferns 45

T
Talbotia elegans 25
Tamatave 11
Tasmania 9
Tea 52, 62
Ternstroemia 24
The classification of ferns 41
Thesaurus capensis 9, 17, 20
Thiselton-Dyer, Sir W. 29, 40, 42, 44, 49, 56, 57, 60, *61*, 61, 63, 67, 71, 73, 74, 76, 80, 81, 83, 85
Thunbergia natalensis 24, 24
Thunberg's Flora Capensis 20
Times 53
Tobacco 52, 86
Tongaat 7
Topham,R. 46, 54, 55, 85
Transkei 80
Transvaal 45, 80
Transvaal Republic xxi, 76
Travels in the interior of South Africa, 15
Trichilia sp. 24
Trichomanes pyxidiferum 41
Trinity College, Dublin 9 60
Tryphia 47
Tsetse fly 75
Tugela 5, 43, 47, 73, 78
Turraea floribunda 83
  heterophylla 83
Tyson, W. 71

U
Uba 69 80
Ubombo 74

Uitenhage 62
Umvoti 53
Union Packet Company 80
Ustilago sacchari 52

V

Veitch of Chelsea 86
Vernonia sutherlandii 17
Verulam 43, 56, 78
Victoria amazonica 15
Victoria County 53
Victoria, Queen, xvii
Victoria water lily *30*, 76
Villabrunea integrifolia 85
    integrifolia 84
Voortrekkers xx

W

Wagon 78
Walker, J. 33
Ward, Dr N. 14, 22
Wardian case 3, 6, 14, 19, *22*, 22, 23, 26, 46, 64, 65, 79, 80, 84, 86
Watson, W. 60
Webbia pinifolia 25
Wehdemann, C.H. 2
Wells-next-the-sea 55
West Indies 9, 61
White, A.S. 33, 46, 47, 83, 85
White House estate xv
Wild tea 62
William IV xvii
Wilson, J. xvi
Wolseley, Gen. 50
Wood, J.M. 41, 42, 45, 52, 55, 58, 62, 66, 67, 70, 71, 73, 74, 78, 79, 81, 82, 83, 84
Wylie, J. 67, 71, 74, 76, 84

X

X club 19
Xanthochymus pictorius 26
Xanthoxylum sp 24

Z

Zamia 46
Zizania aquatica 86
Zulu 4, 6, 8, 55
Zululand 4, 5, 6, 8, 10, 21, 31, 73, 78, 83
Zulus, 19
Zwartkop 14, 44